AF792007

Post-Silicon and Runtime Verification for Modern Processors

Ilya Wagner • Valeria Bertacco

Post-Silicon and Runtime Verification for Modern Processors

Springer

Ilya Wagner
Platform Validation Engineering Group
Intel Corporation
Hillsboro, Oregon
USA
ilya.wagner@intel.com

Valeria Bertacco
Department of Electrical Engineering
and Computer Science
University of Michigan
Ann Arbor, Michigan
USA
valeria@umich.edu

ISBN 978-1-4419-8033-5 e-ISBN 978-1-4419-8034-2
DOI 10.1007/978-1-4419-8034-2
Springer New York Dordrecht Heidelberg London

© Springer Science+Business Media, LLC 2011
All rights reserved. This work may not be translated or copied in whole or in part without the written permission of the publisher (Springer Science+Business Media, LLC, 233 Spring Street, New York, NY 10013, USA), except for brief excerpts in connection with reviews or scholarly analysis. Use in connection with any form of information storage and retrieval, electronic adaptation, computer software, or by similar or dissimilar methodology now known or hereafter developed is forbidden.
The use in this publication of trade names, trademarks, service marks, and similar terms, even if they are not identified as such, is not to be taken as an expression of opinion as to whether or not they are subject to proprietary rights.

Printed on acid-free paper

Springer is part of Springer Science+Business Media (www.springer.com)

To my niece Ellie, who showed me
the miracle of learning.

Ilya Wagner

To all my students, who make working
in the field of verification such a rewarding
experience.

Valeria Bertacco

Preface

The growing complexity of modern processor designs and their shrinking production schedules cause an increasing number of errors to escape into released products. Many of these escaped bugs can have dramatic effects on the security and stability of consumer systems, undermine the image of the manufacturing company and cause substantial financial grief. Moreover, recent trends towards multi-core processor chips, with complex memory subsystems and sometimes non-deterministic communication delays, further exacerbate the problem with more subtle, yet more devastating, escaped bugs. This worsening situation calls for high-efficiency and high-coverage verification methodologies for systems under development, a goal that is unachievable with today's pre-silicon simulation and formal validation solutions. In light of this, functional post-silicon validation and runtime verification are becoming vitally important components of a modern microprocessor development process. Post-silicon validation leverages orders of magnitude performance improvements over pre-silicon simulation while providing very high coverage. Runtime verification solutions augment the hardware with on-chip monitors and checking modules that can detect erroneous executions in systems deployed in the field and recover from them dynamically.

The purpose of this book is to present and discuss the state of the art in post-silicon and runtime verification techniques: two very recent and fast growing trends in the world of microprocessor design and verification. The first part of this book begins with a high-level overview of the various verification activities that a processor is subjected to as it moves through its life-cycle, from architectural conception to silicon deployment. When a chip is being designed, and before early hardware prototypes are manufactured, the verification landscape is dominated by two main groups of techniques: simulation-based validation and formal verification. Simulation solutions leverage a model of the design's structure, often written in specialized hardware programming languages, and validate a design by providing input stimuli to the model and evaluating its responses to those stimuli. Formal techniques, on the other hand, treat a design as a mathematical description of its functionality and focus on proving a wide range of properties of its functional behavior. Unfortunately, these two categories of validation methods are becoming increasingly inadequate in coping with the complexity of modern multi-core systems. This is exactly where post-silicon and runtime validation techniques, the primary scope of this book, can lend a much needed hand.

Throughout the book we present a range of recent solutions in these two domains, designed specifically to identify functional bugs located in different components of a modern processor, from individual computational cores to the memory subsystem and on-chip fabrics for inter-core communication. We transition into the second part of the book by presenting mainstream post-silicon validation and test activities that are currently being deployed in industrial development environments and outline important performance bottlenecks of these techniques. We then present *Reversi*, our proposed methodology to alleviate these bottlenecks in processor cores. Basic principles of inter-core communication through shared memory are overviewed in the following chapter, which also details new approaches to validation of communication invariants in silicon prototypes. We conclude the discussion of functional post-silicon validation with a novel echnique, targeted specifically to modern multi-cores, called *Dacota*.

The recently proposed approaches to validation that we collected in part two of this book have an enormous potential to improve verification performance and coverage; however, there still is a chance that complex and subtle errors evade them and escape into end-user silicon systems. Runtime solutions, the focus of the third part of this work, are designed to address these situations and to guarantee that a processor performs correctly even in presence of escaped design bugs without degrading user experience. To better analyze these techniques we investigate the taxonomy of escaped bugs reported for some of the processor designs available today, and we also classify runtime approaches into two major groups: *checker-* and *patching-based*. In the remainder of part three we detail several runtime verification methods within both categories, first relating to individual cores and then to multi-core systems. We conclude the book with a glance towards the future, discussing modern trends in processor architecture and silicon technology, and their potential impacts on the verification of upcoming designs.

Acknowledgements

We would like to acknowledge several people that made the writing of this book possible. First, and foremost, we express our gratitude to our colleagues, who worked with us on the research presented in this book. In particular, we would like to thank Professor Todd Austin, who was a vital member of our runtime verification research and provided critical advice in many other projects. Andrew DeOrio has contributed immensely to our post-silicon validation research and has helped us greatly in the experimental evaluations of several other techniques. Both Todd and Andrew have also worked tirelessly on the original development of these works and provided valuable insights on the presentation of the material in this manuscript.

We would also like to thank many students working in the Advanced Computer Architecture Lab, and, especially, all those who have devoted their research to hardware verification: Kai-hui Chang, Stephen Plaza, Andrea Pellegrini, Debapriya Chatterjee, Rawan Abdel-Khalek have worked particularly closely to us among many others. Every day these individuals are relentlessly advancing the theory and the practice of this exciting and challenging field. We also acknowledge all of the faculty and staff at the Computer Science and Engineering Department of The University of Michigan, as well as all engineers and researchers in academia and industry who provided us with valuable feedback on our work at conferences and workshops, reviewed and critiqued our papers and published their findings, upon which much of our research was built. These are truly the giants, on whose shoulders we stand.

We also thank our families, who faithfully supported us throughout the years of research that led to the publication of this book. Each and every one of them was constantly ready to offer technical advice or a heartwarming consolation in difficult times, and celebrate the moments of our successes. Indeed, without their trust and encouragement this writing would be absolutely impossible.

Finally, we would like to acknowledge our editors, Mr. Alex Greene, Ms. Ciara Vincent and Ms. Katelyn Chin from Springer, who worked closely with us on this publication with truly angelic patience. Time and again, they encouraged us to continue writing and gave us valuable advice on many aspects of this book.

Contents

Part III RUNTIME VERIFICATION FOR MODERN MICROPROCESSORS

Acronyms

ALU	Arithmetic-logic unit. A hardware block that performs integer arithmetic and logic functions, such as addition, subtraction, logic AND, *etc.*
API	Application programming interface. A set of functions and routines which describe the interface of a software application. API description is typically limited to the application interface, and does not specify the way the functionality of the program is actually implemented.
ATPG	Automatic test pattern generation. A technique for post-silicon validation of a manufactured circuit, which attempts to expose errors in fabrication of individual gates and interconnect in the design. ATPG tools use a software representation of the netlist to derive input stimuli that can expose a variety of manufacturing defects and then apply these tests to the prototype.
BDD	Binary decision diagrams. A data structure for compact representation and fast operations on Boolean logic functions. BDDs are commonly used as an underlying engine of formal verification approaches, such as symbolic simulation, reachability analysis and model checking.
BIOS	Basic input-output system. A firmware residing on the motherboard, that tests and configures various modules of a computer system upon startup.
BMC	Bounded model checking. A pre-silicon verification technique which establishes adherence of design's behavior, within a finite number of clock cycles, to formal specifications, often written as temporal logic formulas.
CPI	Cycles per instruction. A metric of processor performance, which measures the average number of clock cycles that the device needs to perform one operation.
CPU	Central processing unit. Electronic circuit capable of executing software programs, synonymous to the word "microprocessor".
DFT	Design for testability. A class of techniques, which enable testing and debugging of digital circuits, *e.g.,* scan-chains, boundary-scan, on-chip logic analyzers, *etc.*
ECC	Error-correcting code. A special redundant encoding of the data that allows to recover the information even if some of the bits of it are corrupted. Typically used to protect computer storage, such as memory and caches.

- **EDA** Electronic design automation. A generic name for computer-aided tools for electronics design, as well as for the companies producing and deploying such tools.
- **FPGA** Field-programmable gate array. A digital circuit, which can be programmed to implement arbitrary logic functions. FPGAs are commonly used to emulate behavior of complex devices, such as microprocessors, before the first prototype is manufactured.
- **FPU** Floating point unit. A hardware block that performs floating point computation inside of the processor.
- **FRCL** Field-repairable control logic. A hardware patching solution, which augments a processor with a programmable matcher to detect and recover from erroneous control logic states.
- **FSM** Finite state machine. A graph description of hardware block operation. Consists of graph vertices, describing states of the machine, and edges, identifying legal transitions between the states.
- **GPU** Graphics processing unit. An electronic circuit dedicated to processing graphical information, before it is sent to a computer display. GPU chips are commonly placed on the motherboard or implemented as a separate graphics card.
- **HDL** Hardware description language. A class of programming languages that are used to describe functionality and organization of digital hardware, so the behavior of the design can be simulated.
- **ILP** Instruction level parallelism. A potential overlap in the execution of independent instructions. ILP measures how many operations in a program can be performed in parallel.
- **ISA** Instruction set architecture. A complete list of all instructions and operations that a processor can execute. ISA also often includes a complete specification of the processor interface, in terms of communication protocols, interrupt handling, *etc.*
- **JTAG** Joint test action group. Initially, an industry group that designed boundary-scan technique for circuit board testing. Later, term JTAG became synonymous with the boundary scan architecture.
- **NMR** N-modular redundancy. A resiliency technique, where *N* modules (identical or heterogeneous) compute the same function in parallel. Errors in one unit can then be detected and corrected through majority voting scheme.
- **OCLA** On-chip logic analyzer. A circuit, residing on a processor die, which can be programmed to monitor for specific activity of the processor and record the internal state of the chip upon occurrence of the trigger.
- **OS** Operating system. A software supervisor that manages the hardware and user-level programs of a computer system. An operating system provides applications with access to the hardware and coordinates the resource sharing in a computer system.
- **PCI** Peripheral component interconnect. A type of computer bus that connects peripheral devices of a system, *e.g.,* graphics and network cards, to the motherboard and the processor.

PSMI Periodic state management interrupt. A technology developed at Intel, which allows to periodically stop program execution and collect the internal state of a silicon prototype for post-silicon debugging.

QoS Quality of service. A network control mechanism that distributes resources to different classes of traffic to achieve deterministic or statistical guarantees of system performance.

ROB Re-order buffer. A hardware block for instruction re-ordering used in complex out-of-order processor architectures. To hide the effects of long-latency operations a processor may issue instructions in an order different than that of the original program and use an ROB to ensure that the instruction stream is committed in strict program order.

ROM Read-only memory. A class of storage of computer devices, where stored data cannot be modified. In microprocessor domain read-only memory is typically implemented as a hard-wired lookup table.

RTG Random test generator. A software that creates randomized sequences of inputs to the design for testing purposes. Typically the test stimuli are not entirely random, but are constrained to a subset of all valid inputs to the design.

RTL Register transfer level. An register transfer level description of a logic device consists of memory elements (registers) and functions that transfer the data between them. An RTL description is typically implemented in a hardware description language.

SAT Boolean satisfiability. A class of problems that establishes if there exist an assignment to variables of a Boolean formula that evaluates it to true. SAT-solvers and their derivatives are often used by formal verification approaches, such as equivalence checking.

SEU Single event upset. A change or corruption of the state of a latch (or a flip-flop), due to an energetic particle strike.

SPICE Simulation program with integrated circuit emphasis. A class of simulation software that can evaluate the behavior of an integrated circuit from the electrical standpoint. SPICE technique solve complex differential equations, which describe how voltage and current at in various points in the design change over time, thus, operating at lower abstraction level than logic simulation solutions.

STG State transition graph. A graph description of states and transitions between the states that hardware can assume at runtime. Synonymous to finite state machine.

SoC System-on-a-chip. A hardware device that integrates multiple components of a computer system, *e.g.*, processing cores, non-volitile memory, peripheral controllers, *etc.* on a single silicon die.

TLB Translation look-aside buffer. A hardware module that acts as a fast cache to a larger and slower lookup table, thus increasing processor performance for accesses that hit in the TLB.

Part I
VERIFICATION OF A MODERN PROCESSOR

This part of the book provides a high-level overview of the design and verification cycle of a modern processor, starting from the early design phases, to prototype manufacturing and testing and to product release. In the second chapter we explore each of the three major phases of the verification universe: pre-silicon verification, post-silicon validation and runtime techniques. For each phase, we discuss several key solutions and investigate their main advantages and drawbacks. Most importantly, we compare performance and coverage of verification methods across different phases, making the case for functional post-silicon and runtime verification techniques.

Chapter 1
VERIFICATION OF A MODERN PROCESSOR

Abstract. Over the past four decades, microprocessors have come to be a vital and inseparable part of the modern world, becoming the digital brain of numerous electronic devices and gadgets that make today's lifestyle possible. Processors are capable of performing computation at astonishingly high speeds and are extremely integrated, occupying only a few square centimeters of silicon die. However, this computational power comes at a price: the task of verifying a modern microprocessor and guaranteeing the correctness of its operation is increasingly challenging, even for most established processor vendors. To deliver always higher performance to end-users, processor manufacturers are forced to design progressively more complex circuits and employ immense verification teams to eliminate critical design bugs in a timely manner. Unfortunately, too often size doesn't seem to matter in verification, as schedules continue to slip, and microprocessors find their way to the marketplace with design errors. In this chapter we overview the life-cycle of a microprocessors, discuss the challenges of verifying these devices, and show examples of hardware errors that have escaped into production silicon because of insufficient validation an d their impact.

1.1 The Birth of the Microprocessor

Over the past four decades microprocessors have permeated our world, ushering in the digital age and enabling numerous technologies, without which today's life style would be all but impossible. Processors are microscopic circuits printed onto silicon dies and consisting of hundreds of millions of transistors interconnected by wires. What distinguishes microprocessors from other integrated circuits is their ability to execute arbitrary software programs. In other words, processors make digital devices programmable and flexible, so a single device can efficiently perform various operations, depending on the program that is running on it. In our every day activities we encounter and use these tiny devices hundreds of times, often without even realizing it. Processors allow us to untether our phones from the wired network and

I. Wagner and V. Bertacco, *Post-Silicon and Runtime Verification for Modern Processors*,
DOI 10.1007/978-1-4419-8034-2_1, © Springer Science+Business Media, LLC 2011

enable mobile communications, while their counterparts, deployed by phone companies, made communications richer and much more reliable. Processors monitor the health of hospital patients, control airplanes, tally election votes and predict weather. And, of course, they power millions of personal computers of all shapes and sizes, as well as the backbone of the Internet, a vital and inseparable part of modern life. The computational power of these devices grows every year at an astonishing pace: not long ago processors were only capable of executing just a few thousands operations per second, while today they can perform billions of complex computations per second. Finally, in the past few years, hardware design houses have introduced multi-core processors, that is, systems comprising multiple processors (cores) on the single silicon die. These systems, can execute several programs concurrently, thereby multiplying the overall performance delivered to the user.

However, to be so powerful, processors implement extremely complex architectures, making the design and manufacturing of these devices a major challenge for the semiconductor industry. Companies such as Intel, IBM and AMD are forced to dedicate hundreds of engineers for years at a time to continue to advance microprocessor technology and deliver the next generation processors to end-users. Moreover, as these designs grow in complexity, it becomes increasingly harder to verify them and ensure that they operate properly. Design houses report that today verification efforts significantly overweigh design activities, and that they often staff their teams with two verification engineers per designer. Unfortunately, the complexity and the number of features in each new generation of CPUs have quickly outpaced the capabilities of even the largest industrial teams. As a consequence, today it is impossible to provide high-quality verification of microprocessors with traditional means, and products released to the public are becoming less and less reliable. Furthermore, early in the development process engineers must assess the *verifiability* of all the features that they want to introduce into the new product: if proposed features cannot pass high-quality validation on time and within budget, they cannot be deployed in the final product and are removed from the design plan, resulting in a reduced set of capabilities and performance of the new system.

The consequences of this trend of diminishing quality in verification can be dramatic: indeed, the impact of bugs in production microprocessors can range widely from innocuous to devastating, for several reasons. For instance, it is possible for a computer system to become compromised, in terms of safety and security, because of a hardware bug. As a result, a system with a buggy processor becomes vulnerable to security attacks. Attacks of this type could be perpetrated even on systems running completely correct software, since they rely exclusively on underlying hardware flaws [Int04, Kas08]. Moreover, bugs can have a disastrous financial impact on the manufacturing company by triggering a costly recall of faulty hardware, as was the case in a past Intel processor [Mar08]; or causing significant delays in product release, similar to what happened with AMD's Phenom, released in 2007 [Val07]. The impact in both cases is estimated in billions of dollars, due to the large volume of defective components that a functional bug always entails. To prevent devastating errors from seeping into the released designs, a variety of techniques have been devised to detect and correct issues during system's design and manufacturing. Conceptually, these verification approaches can be divided into three families, based on where they intervene in a processor life-cycle: *pre-silicon, post-silicon* and *runtime verification solutions.*

Pre-silicon techniques are heavily deployed in the early stages of a processor's design, before any silicon prototype of the device is available, and can be classified as simulation-based or formal solutions. Simulation-based methods are the most common approaches to locate design errors in microprocessors. Random instruction sequences are generated and fed into a detailed software model, also called a hardware description of a design, results are computed by simulation of this model and then checked for correctness. This approach is used extensively in the industry, yet it suffers from a number of drawbacks. First, the simulation speed of the detailed hardware description is several orders of magnitude slower than the actual processor's performance. Therefore, only relatively short test sequences can be checked in this phase; for instance, it is almost impossible to simulate an operating system boot sequence, or the complete execution of a software application running on the processor. More importantly, simulation-based validation is a non-exhaustive process: the number of configurations and possible behaviors of a modern microprocessor is too large to allow for the system to be fully checked in a reasonable time.

Another family of pre-silicon solutions, formal verification techniques, solves the non-exhaustive nature of simulation, using sophisticated mathematical derivations to reason about a design. If all possible behaviors of the processor could be described with mathematical formulas, then it would be possible to prove the correctness of the device's operation as a theorem. In practice, in the best scenarios it is possible to guarantee that a design will not exhibit a certain erroneous behavior, or that it will never produce a result that differs from a known-correct reference model. The primary drawback of formal techniques, however, is that they are far from capable of dealing with the complexity of modern designs, and thus their use is limited to only a few, small components within the processor.

After a microprocessor is sufficiently verified at the pre-silicon stage, a prototype is manufactured and subjected to *post-silicon validation* where tests can, for the first time, be executed on the actual hardware. The key advantage of post-silicon validation is that its raw performance enables significantly faster verification than pre-silicon software-based simulation, thus it could deliver much better correctness guarantees. Unfortunately, post-silicon validation presents a challenge when it comes to detection and diagnosis of errors because of the limited observability provided by this technique, since at this stage it is impossible to monitor the internal components of the hardware prototype. Therefore, errors cannot be detected until they generate an invalid result, or cause the system to hang. The limited observability leads to extremely involved and time consuming debugging procedures, with the result that today post-silicon validation and debugging has become the single largest cost factor for processor companies such as Intel.

Due to the limitations of pre- and post-silicon verification, and shrinking timelines for product delivery, processor manufacturers have started to accept the fact that bugs do slip into production hardware and thus they are beginning to explore *runtime verification solutions* that can repair a device directly at the customer's site.

"Patching" microprocessor bugs, however, is a non-trivial task, since the functionality of the device is already embodied in the transistors comprising the silicon die, and it cannot be easily modified at this point. To enable in-the-field patching, designers create special processor components dedicated to detecting erroneous behaviors and recovering from them. Runtime verification is currently in its early research stage: a few techniques have been recently proposed by academic research, while problem-specific solutions are starting to appear in commercial products.

1.2 Verification Throughout the Processor Life-cycle

A traditional microprocessor's design and manufacturing flow (shown in Figure 1.1) consists of a series of steps that considers a high-level description of processor operation (*specification*), refines and transforms it, and, finally, implements the specified functionalities on a silicon die. After each step, the design is progressively verified, to ensure that, after all transformation and concretization steps, the behavior of the device still adheres to the original specification. The process starts with a high-level specification of the microprocessor's required characteristics and functionalities, often described in a natural language, and/or diagrams describing its basic structure and how the device should interface to other digital systems. This specification is then converted to an *architectural model* of the device, typically written in a high-level programming language (such as C). This model represents the first formalized reference of the final system's behavior. Implementation in a hardware description language (HDL) can then start. The HDL description of the design describes the operation of individual sub-modules of the processor, as well as their interactions, and is also known as the *register-transfer level* (RTL) model. This RTL model is then verified to establish its equivalence to the architectural model through simulation-based and formal techniques. The outcome of simulation-based tests is compared to those of the known-correct (or "golden") architectural model and discrepancies, indicators of errors, are identified and fixed. In addition to simulation-based techniques, in *pre-silicon verification*, engineers often employ formal methods, which can check correctness of a design using mathematical proofs and can thus guarantee the absence of certain types of errors. Unfortunately, formal methods cannot handle complex RTL models due to their limited scalability, therefore, their usage is limited to a few, small critical blocks.

Once the RTL model is sufficiently validated, designers use synthesis software that maps HDL into individual logic gates, registers and wires, generating a *netlist* of the circuit. Since conversion from an RTL model to a netlist may incur errors, specialized verification solutions intervene again to check that this new transformation is still equivalent to the previous model. Place and route software applications then calculate how individual logic elements in the netlist can be placed on the silicon die to produce a design that fulfills the required characteristics of power, area, delay, *etc.* After placement, the final description of the design is *taped-out*, *i.e.,* sent to a fabrication facility to be manufactured. When the first hardware prototypes become

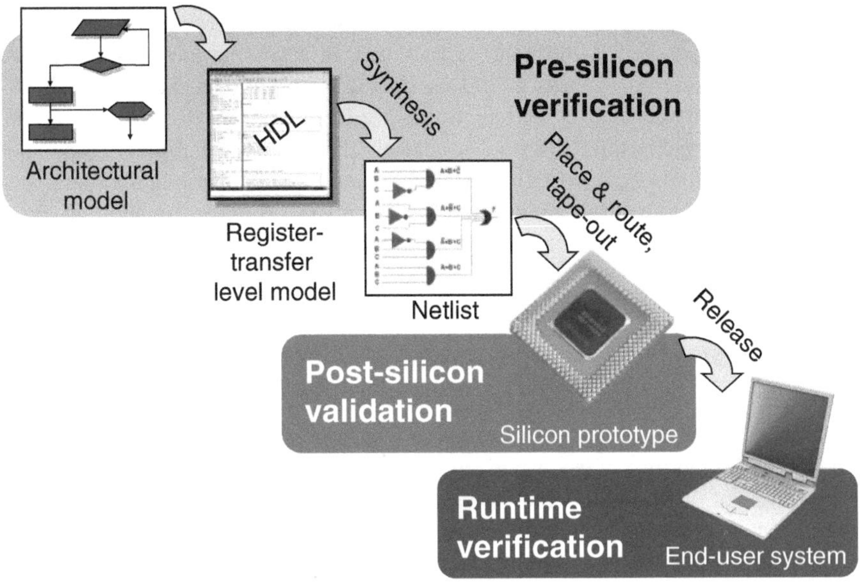

Fig. 1.1 Modern microprocessor design and verification flow. In the pre-silicon phase an architectural model, derived from the original design specification, is converted into an RTL implementation in a hardware description language (HDL). The RTL model can then be synthesized, producing a design netlist. Place and route software calculates where individual logic gates and wire connections should be placed on the silicon die and then a prototype of the processor is manufactured. Once a prototype becomes available, it is subjected to post-silicon validation. Only after the hardware is shown to be sufficiently stable in this validation phase, the processor is released and deployed in the market.

available, they can be inserted into a computer system for *post-silicon validation* (as opposed to the pre-silicon verification that occurs before the tape-out). One of the distinguishing features of post-silicon verification is its high performance: the same performance as the final product, which is orders of magnitude higher than simulation speeds in the pre-silicon domain. Typically, at this stage engineers try to evaluate the hardware in real life-like settings, such as booting operating systems, executing legacy programs, *etc.* The prototype is also subjected to additional random tests, in an attempt to create a diverse pool of scenarios where differences between the hardware and the architectural model can be identified to flag any remaining errors. When a bug is found at this stage, the RTL model is modified to correct the issue and the design often must be manufactured again (this process is called *re-spin*).

A processor design usually goes through several re-spins, as bugs are progressively exposed and fixed in manufactured prototypes. Ultimately, the design is stabilized and it can go into production. Unfortunately, due to the complexity of any modern processor, it is impossible to exhaustively verify its behavior either in pre-silicon or in post-silicon, thus subtle, but sometimes critical, bugs often slip through all validation stages. Until recently, if a critical functional bug was exposed in end-user's hardware, manufacturers had no other choice but to recall the device. Today vendors are starting to develop measures to avoid such costly recalls and allow their products to be patched in the field. Researchers in academia have also proposed solutions to ensure correctness of processor operation with special on-die checkers. Patching- and checker-based techniques are cumulatively classified as *runtime verification* approaches. In Chapter 2 we review the current techniques deployed in pre-silicon, post-silicon and runtime phases in more details, while in the remainder of the book we concentrate on several new promising solutions in the two latter domains.

1.3 Verification of a Modern Processor: a Case Study

The importance and difficulty of microprocessor verification is often underestimated today by casual users and electronic design industry professionals alike. One of the main causes of this is the fact that processor vendors rarely release exact data about their internal validation techniques and experiences into the public domain, and are even more cautions to disclose information about any encountered bugs. This is understandable from the business point of view to a certain degree, since a large portion of the tools and methodology used in validation by design companies are developed in-house and are used as competitive advantage against rival vendors. Likewise, information about any sort of bug, even those that had been fixed before product release, casts a negative image on the manufacturer and could potentially reveal confidential details about the inner-workings of the product. Nevertheless, there are a few publications from processor vendors that shed light on the validation of their products and give some degree of appreciation of the challenges faced by verification engineers.

One of the most comprehensive of such studies reports the pre-silicon validation effort on the Pentium 4 project , which was a new processor designed based on Intel's 7th generation NetBurst architecture [Ben05, Ben01]. In this work, authored by Bob Bentley, the daunting task of verification engineers is vividly described. The design process for Pentium 4 was started at Intel in 1996 and continued until tape-out three years later. During this time the validation team increased from 10 to 70 people, the majority of which had to be trained to use such verification tools as model checking, cluster-level simulation, *etc.* By tape-out time, the processor had undergone a transformation from a high-level idea and an architectural description to an RTL design of more than one million lines and was validated at cluster- and chip-level for more than 200 billions simulation cycles. This number is an astonishing feat if one takes into account that the chip-level simulation speed did not exceed 5Hz, due to sheer size of the RTL code base: that corresponds to more than a thousand years of cumulative simulation time. Consequently, the majority of pre-silicon simulation was done at cluster level, which not only allowed for faster simulation, but added much needed controllability to individual processor blocks. Moreover, as the RTL code was modified, regressions suites were executed time and again to maintain code stability throughout the development. To aid with this verification effort, the engineering team had to resort to large server farms that processed batches of tests around the clock.

Pre-silicon validation, however, was not limited to simulation: the Pentium 4 validation project pioneered an extensive use of formal validation techniques for the microprocessor domain. Yet, those were only applied to a few critical blocks, such as the floating point unit and decoder logic, since the full design, comprising 42 million transistors, was too much for formal tools to handle. Thus, these mathematical approaches were selectively directed towards blocks that had been sources of errors in past designs. The effort payed off well, and several critical bugs were exposed and fixed before tape-out. One of those had a probability of occurrence of 1 in $5 * 10^{20}$ floating-point instructions, thus it was very likely that it would have escaped a simulation-only verification methodology; its discovery overted a disaster that could have been critical for the company.

While the "quality" of bugs caught by formal verification was high, the quantity was fairly low (about 6% of total pre-silicon bugs), since only a few blocks were targeted. Simulation-based validation with randomized tests at cluster-level, on the other hand, provided for the majority of errors exposed (3,411 out of total of 7,855), due to its scalability and relatively high speed. To make randomized test generation as effective as possible, engineers tracked 2.5 million unit-level types of behavior inside the device and directed the testing sequences to cover as many of these conditions as possible. Directed assembly-level tests (over 12,000 in total) also led to a very high error discovery rate, accounting for more than 2,000 bugs. Note that, a number of additional issues were also exposed in testbenches and validation tools, and had to be addressed during the verification process. These issues are not accounted for in the reported 7,855 bugs. Post-silicon validation issues are also not accounted for in the total. This phase of the design cycle of Pentium 4 was only ten months long, yet during this time the device executed orders of magnitude more cycles than in the three years of pre-silicon effort. Operating at speeds of 1GHz and up, these prototypes underwent testing at different temperatures and voltages, run scores of applications and random tests, and communicated with a range of peripheral devices. In addition to time and engineering efforts, post-silicon validation also incurs high equipment costs: verification engineers commonly have to build and debug specialized in-house testing and analysis platforms and purchase test-pattern generators, optical probing machines and logic analyzers, with costs in the range of hundreds of thousands of dollars.

1.4 Looking Ahead

The comprehensive report discussed in the previous section was compiled for a processor designed in the late 90's, nevertheless, it provides an accurate picture of industrial-scale validation, which we can use as a baseline reference for outlining future trends in verification. From the early 1970s, Moore's law implacably has pushed the device density up, and since the release of the first Pentium 4, proces-

Table 1.1 Characteristics of processor designs developed during 2000-2008.

Year	Name	Tech-noglogy (nm)	Number of transistors (mil.)	Die area (mm^2)	Number of cores	Inter-core comm. medium
2000	Intel Pentium 4 Willamette	180	42	217	1	—
2004	Intel Pentium 4 Prescott	90	125	109	1	—
2005	IBM Cell	90	234	221	9	bus
2005	UltraSPARC T1 Niagara	90	300	378	8	crossbar
2006	Intel Core 2 Merom	65	291	143	2	bus
2007	AMD Phenom X4 Agena	65	463	285	4	bus
2007	Intel Polaris prototype	65	100	275	80	2D mesh
2007	Tilera TILE64	90	615	430	64	2D mesh
2008	Intel Core i7 Bloomfield	45	731	263	4	crossbar

sor complexity has increased by at least an order of magnitude. In Table 1.1 we report the characteristics of several notable processor designs developed in the past decade for comparison, analysis, and as an indicator of future trends. The first critical point to note is the fact that, since the first Pentium 4, silicon fabrication technology has shrunk from 180 to 45nm. Moreover, various advanced features, such as support for simultaneous multi-threading, multiple power savings techniques and virtualization, have become commonplace. More importantly, modern designs often contain multiple computing cores, sometimes heterogeneous, communicating with each other through on-die media. This diversification of on-chip resources is expected to continue well into the future: designers have already began to integrate peripheral components, such as memory controllers and graphics, into the main processor. Thus, we can foresee that the CPU of the 21st century will be comparable to a complex system on a chip (SoC), with a wide range of functionalities and interdependent components, all requiring thorough validation. Production timelines of these integrated circuits, however, are not expected to increase, since end-users continue to demand higher performance and broader functionality at the same pace. Consequently, designers will be forced to staff larger validation teams and increase investments into simulation servers, prototyping tools, *etc.* Worse still, the performance of traditional verification and design tools in the future will be lagging behind the complexity of final silicon products even more. For instance, due to an increased number of on-die components, the performance of a full chip-level simulation for future designs is not expected to be greater (and will probably be worse) than that of the Pentium 4, despite significant improvements in the simulation hardware hosts. As the number of features grows, in some cases full-system simulation may become unfeasible. Likewise, increasing capabilities of formal verification tools in the future will be outpaced by the complexity of critical modules requiring formal analysis.

In this worsening situation, the number of total bugs in processor products and the speed with which they are discovered in the field is rapidly increasing. Researchers have already reported that the escape bug discovery rate in Core 2 Duo designs is 3 times larger than that of the Pentium 4 design [CMA08]. Note that these are functional errors that have evaded all validation efforts and found their way into the final product, deployed in millions of computer systems world-wide. Thus, they entail a very high risk of having a critical impact on the user base and the design house. It is already clear that because of the expanding gap between complexity and verification effort, in the future errors will continue to slip into silicon, potentially causing much more damage than the infamous FDIV bug, which resulted in a $420 million loss for Intel in the mid-90s [Mar08]. For instance, as recently as 2007, an error in the translation look-aside buffer of the third level cache in the Phenom processor by AMD forced the manufacturer to delay the market release by several months. Not only this delayed the distribution of the product to the market, but also created much negative publicity for the company, and influenced the price of its stock [Val07]. From this grim picture we draw the conclusion that new validation solutions are critically needed to enable the continued evolution of microprocessors designs in the future. This concern is also voiced in the International Technology Roadmap (ITRS) for Semiconductors, which states that "without major breakthroughs, verification will be a non-scalable, show-stopping barrier to further progress in the semiconductor industry." ITRS also reports that there are no solutions available to provide high-quality verification of integrated circuits and sufficiently low rate of escapes beyond the year 2016 [ITR07].

1.5 Summary

Microprocessors have entered our world four decades ago and since that time have played a vital role in our everyday life. Throughout these forty years, the capabilities of these amazing devices have been increasing exponentially, and today processors are capable of performing billions of computations per second, and executing multiple software applications in parallel. This performance level was enabled by rapidly growing complexity of integrated circuits, which, in turn, exacerbated the problem of their verification. To validate modern processors, consisting of hundreds of millions of transistors, manufacturing companies are forced to employ large verification teams, develop new validation technology and invest into costly testing and analysis equipment. Traditionally, this verification activity is broken down into three major steps: pre-silicon verification, post-silicon validation, and runtime verification.The former two are conducted internally by the vendors on a software model of the device and and a silicon prototype, respectively, and are primarily carried out through execution of test sequences. However, as we discussed in this chapter, the complexity of modern processors prohibits their exhaustive verification, and thus errors often slip into final products, causing damage to both end-users and vendor companies. To protect hardware deployed in the field from such errors, researches have recently started to propose techniques for runtime verification and to rethink post-silicon validation strategies to derive better quality of results from them. Some of the emerging techniques in these domains will be presented in detail in the later chapters of this book, however, first, we must take a deeper look into a traditional processor verification cycle to understand advantages and limitations of each of its steps.

References

[Ben01] Bob Bentley. Validating the Intel® Pentium® 4 microprocessor. In *DAC, Proceedings of the Design Automation Conference*, pages 224–228, June 2001.

[Ben05] Bob Bentley. Validating a modern microprocessor. In *CAV, Proceedings of the International Conference on Computer Aided Verification*, pages 2–4, July 2005.

[CMA08] Kypros Constantinides, Onur Mutlu, and Todd Austin. Online design bug detection: RTL analysis, flexible mechanisms, and evaluation. In *MICRO, Proceedings of the International Symposium on Microarchitecture*, pages 282–293, November 2008.

[Int04] Intel Corporation. *Intel®Pentium®Processor Invalid Instruction Erratum Overview*, July 2004. `http://www.intel.com/support/processors/pentium/sb/cs-013151.htm`.

[ITR07] International Technology Roadmap for Semiconductors executive summary, 2007. `http://www.itrs.net/Links/2007ITRS/Home2007.htm`.

[Kas08] Kris Kaspersky. Remote code execution through Intel CPU bugs. In *HITB, Proceedings of Hack In The Box Conference*, October 2008.

[Mar08] John Markoff. Burned once, Intel prepares new chip fortified by constant tests. *New York Times*, November 2008. `http://www.nytimes.com/2008/11/17/technology/companies/17chip.html`.

[Val07] Theo Valich. AMD delays Phenom 2.4 GHz due to TLB errata. *The Inquirer*, November 2007. `http://www.theinquirer.net/inquirer/news/995/1025995/amd-delays-phenom-ghz-due-tlb`.

Chapter 2
THE VERIFICATION UNIVERSE

Abstract. In this chapter we take the reader through a typical microprocessor's life-cycle, from its first high-level specification to a finished product deployed in a end-user's system, and overview the verification techniques that are applied at each step of this flow. We first discuss pre-silicon verification, the process of validating a model of the processor at various levels of abstraction, from an architectural specification to a gate-level netlist. Throughout the pre-silicon phase, two main families of techniques are commonly used: formal methods and simulation-based solutions. While the former provide mathematical guarantees of design correctness, the latter are significantly more scalable and, consequently, are more commonly used in the industry today. After the first few prototypes of a processor are manufactured, validation enters the post-silicon domain, where tests can run on the actual silicon hardware. The raw performance of in-hardware execution is one of the major advantages of post-silicon validation, while lack of internal observability and limited debuggability are its main drawbacks. To alleviate this, designers often augment their creations with special features for silicon state acquisition, which we review here. After an arduous process of pre- and post-silicon validation, the device is released to the market and finds its way into a final system. Yet, it may still contain subtle bugs, which could not be exposed earlier by designers due to very compressed production timelines. To combat these *escaped errors*, vendors and researchers in industry and academia have began investigating alternative dynamic verification techniques: with minimal impact on the processor's performance, these solutions monitor its health and invoke specialized correction mechanisms when errors manifest at runtime. As we show in this chapter, all three phases of verification, pre-silicon, post-silicon and runtime, have their unique advantages and limitations, which must be taken into account by design houses to attain sufficient verification coverage within their time and cost budgets and to avoid major catastrophes caused by releasing faulty processor products to the commercial market.

I. Wagner and V. Bertacco, *Post-Silicon and Runtime Verification for Modern Processors*,
DOI 10.1007/978-1-4419-8034-2_2, © Springer Science+Business Media, LLC 2011

2.1 Pre-silicon Verification

Pre-silicon verification is a multi-step process, which is aimed at establishing if a design adheres to its specification and fulfills the designer's intentions. Pre-silicon verification, or design-time verification, is conducted before any silicon prototype is available (hence its name), and operates over a range of different descriptions of a digital design: architectural, RTL, gate-level, *etc.* The higher levels of abstraction of the design allow engineers to check, often through mathematical proofs, fundamental properties of a circuit's operation, such as absence of erroneous behaviors and adherence to formally specified invariants in its functionality. The design is then progressively refined to include more detail, requiring the designers to check correctness after each transformation. However, with the growing level of detail, the time and computational effort required to validate the design's functionality increases as well, thus, in practice, only the most critical blocks of a modern processor are fully verified at all levels of abstraction before tape-out.

The history of pre-silicon verification of digital circuits goes hand in hand with the evolution of these devices over the last several decades: what began as a fairly simple in-house activity, quickly became a large industry of its own. Today, tens of electronic design automation (EDA) companies and a handful of the largest digital design houses offer designers a wide variety of software and hardware tools to address various phases of pre-silicon verification; while researchers in industry and academia alike publish thousands of papers yearly on the subject. Some of these techniques can be applied to any logic design, while others are domain-specific heuristics that improve the performance of verification for certain types of designs. Since it would be impossible for us to cover the entire spectrum of solutions available, in this section we simply focus on presenting a concise overview of the most vital verification techniques, and provide a range of references at the end of the chapter for the readers interested in deepening their knowledge on the subject.

2.1.1 From specification to microarchitectural description

In today's microprocessor industry, verification begins early in the design cycle, when architects of the new design develop a detailed set of specifications, outlining the major component blocks and describing their functionality. This specification document is then used as the yardstick to verify various pre-production implementations of the design at varying levels of detail. Note, however, that the document is typically written by several people in a natural human language, such as English, and, therefore, may contain hidden ambiguities or contradictions. Consequently, when bugs are found during the verification of a behavioral model, these might be due to a poor implementation or may be caused by errors in the specification itself, which then must be clarified and updated. As soon as a design specification becomes available, verification engineers begin crafting a *test plan* that outlines how individual blocks and the entire system will be verified, how bugs will be diagnosed and tracked, and how verification thoroughness, also known as *coverage*, will be measured. As with the specification, the test plan is not created once and forever, but rather evolves and morphs, as designers add, modify and refine processor's features.

At the same time, with specification at hand, engineers may begin implementing the design at the architectural (or ISA) level, describing the way the processor interacts with the rest of the computer system. For instance, they determine the format of all supported instructions, interrupts, communication protocols, *etc.*, but do not concern themselves with the details of the inner structure of the design. This is somewhat similar to devising the API of a software application without detailed descriptions of individual functions. Note that operating systems, user-level programs and many hardware components in a computing platform interact with the processor at the ISA level, and for the most part, need not be aware of its internals. In the end, the specification is transformed into an *architectural simulator* of the processor, which is typically a program written in a high-level programming language. As an example, Simics [MCE+02] and Bochs [Boc07] are fairly popular architectural simulators. An architectural simulator enables the engineers to evaluate the high level functionality of the design by simulating applications and it also provides early estimations of its performance. The latter, however, are very approximate, since at this point the exact latency of various operations is unknown. More importantly, at this stage of the development, the architectural simulator becomes the embodiment of the specification and in later verification stages it is referred to as a *golden model* of the design, to which detailed implementations (netlist-level and circuit-level) can be efficiently compared. The architectural simulator can then be refined into a *microarchitectural description* , where the detailed functionality of individual sub-modules of the processor is specified (examples of microarchitectural simulators are SimpleScalar [BA08] and GEMS [MSB+05]). With a microarchitectural simulator designers define the internal behavior of the processor and characterize performance-oriented features such as pipelining, out-of-order execution, branch prediction, *etc.* The microarchitectural description, however, does not specify how these blocks will be implemented in silicon and is usually also written in a high-level language, such as C, C++ or SystemC. Nevertheless, with the detailed information on instruction latency and throughput now available, the performance of the device can be benchmarked through simulation of software applications and other tests.

2.1.2 Verification through logic simulation

The microarchitectural level design is then further refined into a register-transfer level (RTL) implementation, typically written in a hardware design language (HDL), such as Verilog [IEE01b], SystemVerilog [IEE07] or VHDL[IEE04] . At this level, the functionality of individual blocks is further broken down into logic/mathematical operations and storage elements to contain the results of computation. Because of this refinement, the size of the code base becomes significantly larger, while the simulation performance of the design decreases by 4-5 orders of magnitude, compared to the architectural level. With a register-transfer level description it is possible to simulate and evaluate the behavior of the system very accurately, tracking the interaction of all components. The vast majority of the verification effort is dedicated to testing and simulating this level of the design description and most functional design bugs are exposed and corrected at this stage. However, one of the main drawbacks of simulation-based techniques is the fact that they can only identify the presence of errors but cannot guarantee their absence. In other words, with logic simulation designers can check if the processor behaves properly, *i.e.*, adheres to its specification, when running specific program sequences, designed and selected by the design team. However, there still may be latent bugs manifesting only when running other programs, that had not been simulated and verified. Nevertheless, this technique remains the primary method to validate RTL designs, especially in the microprocessor industry, due to its ability to scale and operate even on very complex processor descriptions. During logic simulation, internal design signals are monitored and evaluated one by one, making the time required to complete the simulation directly proportional to the size of the design and the length of the executed test sequence. Therefore, full processor designs can be simulated for millions of cycles, often providing satisfactory confidence that the system is error-free. Example of commercial logic simulators available today include VCS [Syn09] and ModelSim [Men08]. In this chapter we overview the basic components of a logic simulation framework, while the reader is encouraged to investigate surveys, such as [GTKS05, Mei93], and works on functional verification, such as [Mic03, WGR05], for a more detailed analysis of a range of simulation tools, setups and techniques.

In a typical framework for RTL simulation, illustrated in Figure 2.1, the design under validation is wrapped in a test environment that models the behavior of other components external to the system under verification. For instance, when the entire processor is simulated, the test environment represents systems external to the CPU, such as memory, peripherals, *etc.* The test environment and the design are executed together within the logic simulator. The simulator itself is a software application parsing the RTL code of the design and the test environment and computing circuit's output values when subjected to a given test. The outputs are generated by evaluating each internal design component throughout the entire simulated time interval. A simulator obtains the values of primary inputs to the design from a test and then evaluates the circuit's internal logic, as well as its primary outputs, in discrete timesteps. The obtained data can then be saved into a file for later comparison with the outputs of an architectural simulator, used as a golden model, or can be presented to a verification engineer in form of *waveforms*, *i.e.*, a diagram of the signals' values over time. The latter form is often adopted when diagnosing incorrect test responses to investigate the root cause of a bug, since all signal transitions can be visualized.

Test sequences supplied to the simulator can be deterministic, such as manually-generated assembly tests, or may be pseudo-random. The former are usually short and designed to validate specific features or modules of the device, as required by the test plan. Randomized tests are most commonly focused on system-level aspects and interactions, subjecting the design to a variety of stressful stimuli. To this end, verification engineers leverage *pseudo-random test generators* (also called RTGs) that can be tuned to produce streams of valid instructions with a wide range of properties, *e.g.*, focused activity on certain functional units, specific inter-instruction dependencies, and so on [BLL$^+$04, AAF$^+$04]. Some RTGs have the ability to monitor a number of pre-selected internal design signals and use the information to dynamically adjust the test they generate so to boost stimulus quality [WBA07, FZ03]. Furthermore, the most relevant random and directed tests are combined into *regression suites*: sets of tests simulated each time that the design is modified, to guarantee that that the changes did not produce new bugs.

As mentioned before, simulation-based techniques can not provide guarantees of design correctness for the scenarios that are not explicitly tested. To gauge the need for additional verification throughout an industrial-scale digital system development, designers track coverage, which is a measure of thoroughness of verification. For example, achieving a 100% *code coverage* is a typical goal, thus requiring that every statement in the RTL code is activated in simulation least once. Code coverage alone is a fairly weak metric to evaluate the completeness of a test suite; thus, more sophisticated measures such as *functional* and *transaction coverage* are commonly used in addition to code coverage [Glu06]. As discussed in detail in [Piz04], functional coverage metrics allow to evaluate the testing quality of high-level functions in a design's block in a thorough and systematic fashion.

One of the shortcomings of the framework just described is its reliance on the architectural simulator to detect errors. Because modern microprocessors are extremely complex systems, with vast amounts of internal state, subtle errors in inner blocks can often take hundreds or thousands of cycles to manifest at the architectural level. Tracing an error backwards from an erroneous architectural event to the actual malfunction is a tedious and time-consuming process; therefore, the ability to detect issues as soon as they occur is a major advantage in terms of length of the diagnosis effort. One way to achieve this, is to use high-level behavioral models of individual blocks, instead of a monolithic architectural simulator. These models can be derived from the microarchitectural description of the processor and converted to dedicated checking units, running concurrently with the RTL simulation and comparing their output to the corresponding block's simulated output. Sometimes, block-level errors can be identified with simple scoreboards, which could, for instance, keep track of request-response pairs and thus detect rogue messages. Alternatively, designers can use *assertions* or *checkers* to encode invariants of the design's operation and make sure that they are upheld throughout the simulation. More information on this topic can also be found in [FKL04, YPA06]. Assertions on input signals are also helpful in detecting improper communication between blocks or errors arising from misinterpretation of the specification. Digital logic designers may interpret the specification in different ways while developing their components. However, if they encode their assumptions on the behavior of the surrounding components in the form of assertions, mismatches can be detected early in the development process, as soon as blocks are integrated. Finally, assertions may be used to evaluate coverage by monitoring the input signals into the assertion. To enable these advanced verification concepts, in the past few years, the electronic design industry has developed

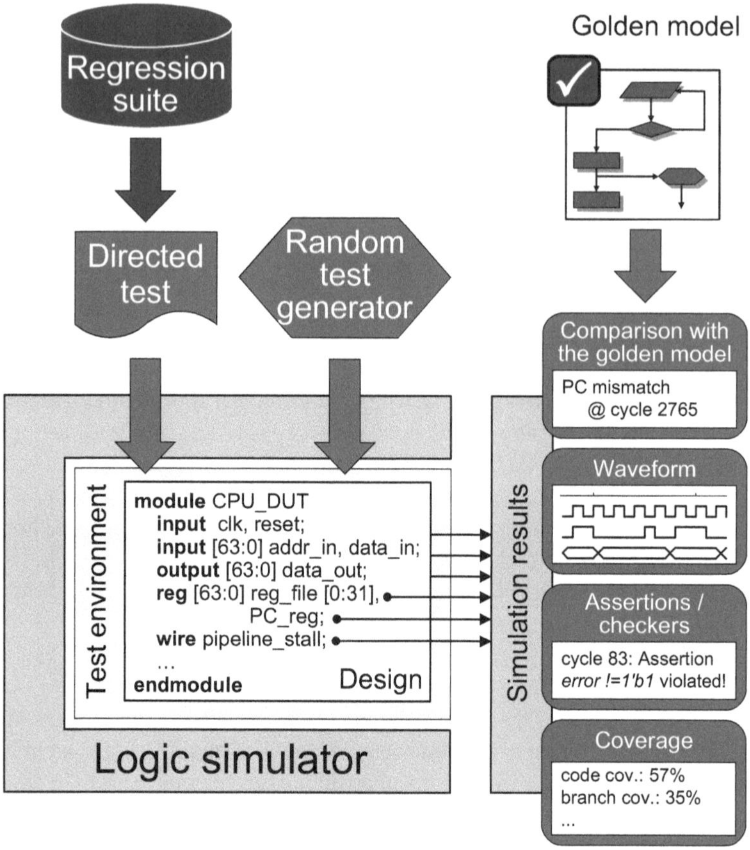

Fig. 2.1 A typical framework for simulation-based verification. Inputs to a logic simulator are typically manually-written directed tests and/or randomized sequences produced automatically by a pseudo-random test generator. Tests are fed into a logic simulator, which, in turn, uses them as stimuli to the integrated test environment and design description. The test environment emulates the behavior of blocks surrounding the design under test. The simulator computed outputs and internal signal values of the design from the test's inputs. These outputs can then be analyzed with a variety of tools: they can be compared with the output of a golden architectural model, can be viewed as waveforms, particularly for error diagnosis, can be monitored by assertions and checkers to detect violations of invariants in the design behavior and, finally, can be used to track coverage, thus evaluating the thoroughness of the test.

dedicated languages for hardware verification (HVLs), some examples of which are OpenVera [HKM01], the *e* language [HMN01] and SystemVerilog [IEE07]. The latter implements a unified framework for both hardware design and verification, providing an object-oriented programming model, a rich assertion specification language, and even automatic coverage collection.

2.1.3 Formal verification

Formal verification encompasses a variety techniques that have one common principle: to prove with mathematical guarantees that the design abides to a certain specification under all valid input sequences. If the design does not adheres to the specification the formal techniques should produce a counterexample execution sequence, which describes one way of violating the specification. To this end, the design under verification and the specification must be represented as mathematical/logic formulas subjected to formal analysis. One of the most important advantages of these solutions over simulation-based techniques is the ability to prove correctness for *all legal stimuli sequences*. As we described in the previous section, only explicitly simulated behaviors of the design can be tested for correctness and, consequently, only the presence of bugs can be established. Formal verification approaches, however, can reason about absence of errors in a design, without the need to exhaustively check all of its behaviors one by one. The body of work in the field of formal verification is immense and diverse: for decades researchers in industry and academia have been developing several families of solutions and algorithms, which are all too numerous to be fully discussed in this book. Fortunately for the readers, sources such as [Ber05, PF05, PBG05, BAWR07, WGR05, CGP08, KG99] and many others, describe the solutions in this domain in much greater detail. In this book, we overview some of the most notable techniques in the field in the hope to stir the readers' curiosity to investigate this research field further. However, before we begin this survey, we first must take a look at two main computation engines, which empower a large fraction of formal methods, namely SAT solvers and binary decision diagrams (BDDs).

SAT is a short-hand notation for Boolean satisfiability, which is the classic theoretical computer science NP-complete problem of determining if there exists an assignment of Boolean variables that evaluates a given Boolean formula to *true*, or showing that no such assignment exists. Therefore, given Boolean formulas describing a logic design and a property to be verified, a SAT-based algorithm constructs an instance of a SAT problem, in which a satisfactory assignment of variables represents a violation of the property. For example, given a design that arbitrates bus accesses between two masters, one can use a SAT-solver to prove that it will never assert both grant lines at the same time. Boolean satisfiability can also be applied to *sequential circuits*, which are in this case "unrolled" into a larger design by replication of the combinational logic part and elimination of the internal state elements. Engineers can then check if certain erroneous states can be reached in this "unrolled" design or if invariants of execution are always satisfied. As we will describe in the subsequent section, SAT techniques can also be used for equivalence checking, *i.e.,* establishing if two representations of a circuit behave the same way for all input sequences. Today, there is a variety of stand-alone SAT-solver applications available, typically they either implement a variation of the Davis-Putnam algorithm [DP60] as, for instance, GRASP [MSS99] and MiniSAT [ES03], or use stochastic methods to search for satisfiable variable assignments as in WalkSAT [SKC95]. Heuristics and inference procedures used in these engines can dramatically improve the performance of the solver; however, in the worst case the satisfiability problem remains NP, that is, as of today, it requires exponential time on the input size to complete execution. As a result, solutions based on SAT solvers cannot be guaranteed to provide results within reasonable time. Furthermore, when handling sequential designs, these techniques tend to dramatically increase the size of the SAT problem, due to the aforementioned circuit "unrolling", further exacerbating the problem.

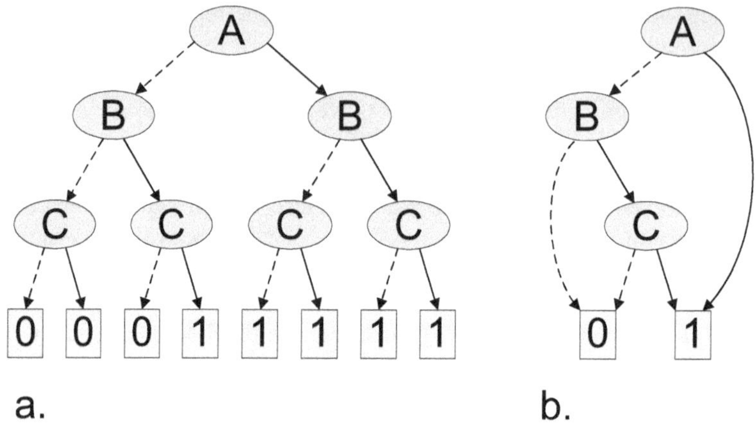

Fig. 2.2 Binary decision diagrams. a. A full-size decision tree for the logic function $f = A|(B\&C)$: each layer of nodes represents a logic variable and edges represent assignments to the variable. In the figure, solid edges represent a 1 assignment, while dashed edges represent a 0 value of the variable. The leaf nodes of the tree are either of 1- or 0-type and represents the value that the entire Boolean expression for the assignment in the corresponding path from root to leaf. Note that the size of the full-size decision tree is exponential in the number of variables in the formula. **b.** Reduced ordered binary decision diagram for the logic function $f = A|(B\&C)$. In this data structure redundant nodes are removed, creating a compact representation for the same Boolean expression.

Binary decision diagrams (BDDs), the second main computational engine used in formal verification, are data structures to represent Boolean functions [Bry86]. BDDs are acyclic directed graphs: each node represents one variable in the formula and has two outgoing edges, one for each possible variable assignment – 0 and 1 (see Figure 2.2). There are two types of terminal (also called "leaf") nodes in the graph, 0 and 1, which correspond to the value assumed by the entire function for a given variable assignment. Thus, a path from the root of the graph to a leaf node corresponds to a variable assignment. The corresponding leaf node at the end of the path represents the value that the logic function assumes for the assignment. BDDs are reduced to contain fewer nodes and edges and, as a result, can often represent complex Boolean functions compactly [Bry86, BRB90]. An example of this is illustrated in Figure 2.2.b, where redundant nodes and edges in the complete binary decision tree for the function $f = A|(B\&C)$ are removed, creating a structure that is linear in the number of variables. BDD software packages as, for instance CUDD [Som09], contain routines that allow a fast and efficient manipulation of BDDs and on-the-fly tree minimization algorithms. Within formal verification, binary decision diagrams can be used for equivalence checking, *i.e.*, testing if two circuits implement the same logical function, in reachability analysis and symbolic simulation, as well as in other techniques. In reachability analysis, BDDs are used to represent the set of states that a design can attain under all possible inputs. This set can be later analyzed to detect the presence of erroneous states. In symbolic simulation, on the other hand, decision diagrams store formulas representing the design's state and the functionality of the outputs in terms of all possible input values. There are, however, a few drawbacks to BDD-based solutions, the most important one being the "memory explosion problem": a situation, where functions cannot be represented compactly by the data structure, causing a prohibitively large number of nodes to be created in the host computer's memory. As a consequence, the host may terminate before completion of the formal proof. The engineering team is then forced to restructure the design or apply simplifying assumptions or *constraints* to the system, abstracting away some of the possible behaviors.

As we already mentioned above, SAT-solvers and BDD libraries enable designers to establish how circuits operate under all possible input conditions and conduct a variety of analyses. Let us now briefly review some of the formal verification solutions that are employed in today's semiconductor industry, outlining their advantages and limitations. Note that only major families of solutions are presented here, since a comprehensive survey is beyond the scope of this book.

Theorem proving is a powerful technique that uses automatic reasoning to derive proof for "theorems" about the system under study. Theorems are mathematical formulas that describe the design's functionality and its properties in abstract form. Theorem provers may use a variety of theories to derive the mathematical transformations required to prove a given theorem. More information may be found in [KK94], which also provides a collection of references on this subject. Theorem provers have found wide acceptance in both the software and hardware verification domains, however, these systems are still not fully automatic and often human assistance is required in directing the proof derivation process. Consequently, engineers spend valuable time investigating conjectures produced by the prover tool and assisting it in reaching its goal. Furthermore, the behavior of the design and the properties are usually specified in abstract form, requiring even more engineering effort. As the result, theorem provers are often applied to high-level protocol or architectural descriptions to ensure that such invariants as absence of deadlock and fairness are upheld. The RTL implementation can then be compared to the formally verified architectural model using more scalable techniques, such as simulation.

The **reachability analysis** problem has been mentioned before and consists of characterizing the set of states that a design can attain during execution. To this end the combinational function of the design is typically represented as a binary decision diagram and the reset state of the circuit is used to initialize the "reached set". Then BDD manipulation functions are used to compute all states that the system can attain after one clock cycle of execution, which are then added to the reached set.

The process continues until the host runs out of memory or the reached set stops increasing, in which case the obtained set contains all states that a system can achieve throughout the execution and can be subsequently analyzed for absence of errors and presence of runtime invariants. Reachability analysis is a very powerful verification tool, although one must keep in mind that this technique, as other BDD-based solutions, suffers from the memory explosion problem and has limited scalability. See also publications such as [CBM89, RS95] for more information on this approach.

Model checking, described in detail in [CGP08, CBRZ01], is another powerful technique that establishes if states and transitions within the design adhere to a formal specification. Properties in this technique are typically written as formulas in temporal logic, which reasons about events in time, for instance, a property may require that a given system's event will eventually occur, independently of the type of execution. However, frequently, formal properties have a limited time window in which they need to be considered, in which case bounded model checking (BMC) can be used. In bounded model checking a property is considered only over a finite number of clock cycles of execution: if no violation is exposed then the property holds. Solutions in this domain may use either BDDs or SAT solvers as the underlying engine. A variety of property specification languages exists today, including PSL [Acc04], recently extended to PSL/Sugar [CVK04], System Verilog Assertion language (SVA) [VR05], which is a subset of SystemVerilog, the proprietary Bluespec language [Arv03], TLA+ [Lam02], *etc.* However, since these languages are declarative, they are often fairly complex to use in describing non-trivial properties and in general are more difficult to use than imperative languages. Indeed, industry experts report that, when a property violation is exposed in model checking, it is more often the case that the problem lies in the property description than in the design itself. The scalability of model checking may also be limited due to the underlying engines's complexity and memory demands.

Symbolic simulation [CJB79, Bry85, BBB$^+$87, BBS90, Ber05] has many similarities with logic simulation, in that output functions are computed based on input values. The main difference here is that inputs are now symbolic Boolean variables and, consequently, outputs are Boolean functions. For instance, the symbolic simulation of the small logic block in Figure 2.3 produces the Boolean expression $A\&(B|C)$ for the output, which we show in the BDD depicted by the output wire. Expressions for the state of the design and the values of its outputs are updated at each simulation cycle. The Boolean expressions altogether are a compact representation of all the possible behaviors that the design can manifest, for all possible inputs, within the simulated cycles. To find the response of the circuit for a concrete input sequence (a sequence of 0s and 1s), one simply needs to evaluate the computed expressions with appropriate values for the primary inputs. Consequently, symbolic simulation can be used to compute a reached state set (within a fixed number of simulation cycles) or prove a bounded time property. As with other solutions, however, the reliance on BDDs brings the possibility of the simulator exhausting the memory resources of the host before errors in the circuit can be identified.

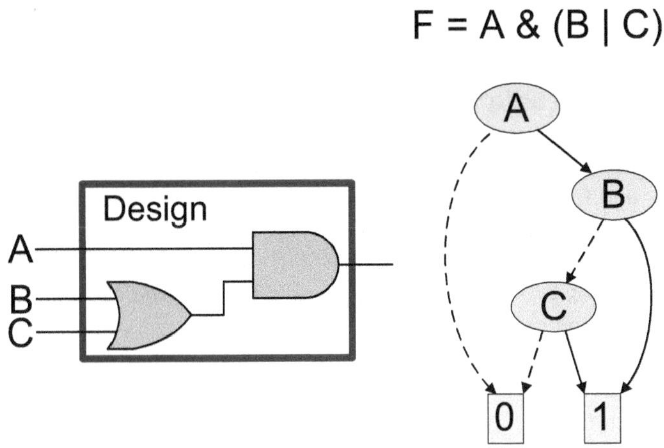

Fig. 2.3 An example of symbolic simulation. In symbolic simulation, outputs and internal design nodes are expressed as Boolean functions of the primary inputs, which, in turn, are described by Boolean variables. In the circuit shown in the figure, symbols A, B and C are applied to the primary inputs and the output assumes the resulting expression $A\&(B|C)$. Functions in symbolic simulation are typically represented by binary decision diagrams, as the one shown at the circuit's output.

As discussed above, modern digital logic designers have access to a wide variety of high-quality formal tools that can be used for different types of analyses and proofs. Yet all of them have limitations, requiring either deep knowledge of declarative languages for specification of formal properties or being prone to exhausting memory and time resources. Consequently, processor designers, who work with extremely large and complex system descriptions, must continue to rely mostly on logic simulation for verification of their products at the pre-silicon level. The guarantees for correctness that formal techniques provide, however, have not been overlooked by the microprocessor industry, and formal tools are often deployed to verify the most critical blocks in modern processors, particularly control units. In addition, researchers have designed methods to merge the power of formal analysis with the scalability of simulation-based techniques, creating *hybrid or semi-formal solutions*, such as the ones surveyed in [BAWR07]. Hybrid solutions use a variety of techniques to leverage the scalability of simulation tools and the depth of analysis of formal techniques to reach better coverage quality with a manageable amount of resources. They often use a tight integration among several tools and make an effort to have them exchange relevant information. Although hybrid solutions combine the best of the formal and simulation worlds, their performance is still outpaced by the growing complexity of processor designs and shortening production schedules. Thus, the verification gap continues to grow and designs go into the silicon prototype stage with latent bugs.

2.1.4 Logic optimization and equivalence verification

The RTL simulation and formal analysis described above are, perhaps, the most important and effort-consuming steps in the entire processor design flow. After these steps are completed, the register-transfer level description of the design must be transformed and further refined before a prototype can be manufactured. The next step in the design flow is *logic synthesis*, which automatically converts an RTL description into a gate-level netlist (see [HS06] for more details of logic synthesis algorithms and tools). To this end, expressions in the RTL source code are expanded and mapped to structures of logic gates, based on the target library specific to the manufacturing process to be used. These, in turn, are further transformed through local and global optimizations in order to attain a design with better characteristics, such as lower timing delay, smaller area, *etc.* Again, with the increasing level of detail, the code base grows further and, correspondingly, the performance of the circuit's simulation is reduced. Formal techniques are often deployed in this stage to check that transformations and optimizations applied during synthesis do not alter the functionality of the design. This task is called *equivalence checking* and it is most often deployed to prove the equivalence of the combinational portion of a circuit. In contrast, sequential equivalence checking has not yet reached sufficient robustness to be commonly deployed in the industry. The basis of combinational equivalence checking relies on the construction of a *miter* circuit connecting two netlists from before and after a given transformation. The corresponding primary inputs in both netlists are tied together, while the corresponding outputs are connected through *xor* gates, as shown in Figure 2.4. Several techniques can then be used to prove that the miter circuit outputs 0 under all possible input stimuli, thus indicating that the outputs of the two netlist are always identical when subjected to identical input. Among these, are BDD-based techniques, used in a fashion similar to symbolic simulation, SAT solvers proving that the miter circuit cannot be satisfied (that is, can never evaluate to 1), and also graph isomorphism algorithms, which allow to prune the size of the circuits whose equivalence must be proven.

After the netlist is synthesized, optimized and validated, it must undergo the *place and route* phase, where a specific location on silicon is identified for each logic gate in the design and all connecting wires are routed among the placed gates [AMS08]. Only after this phase the processor is finally described at the level of individual transistors, and thus engineers can validate properties such as electrical drive strength and timing with accurate SPICE (Simulation Program with Integrated Circuit Emphasis) techniques. It is typical at this stage to only be able to analyze a small portion of the design at a time, and to focus mostly to the critical paths through the processor's sub-modules.

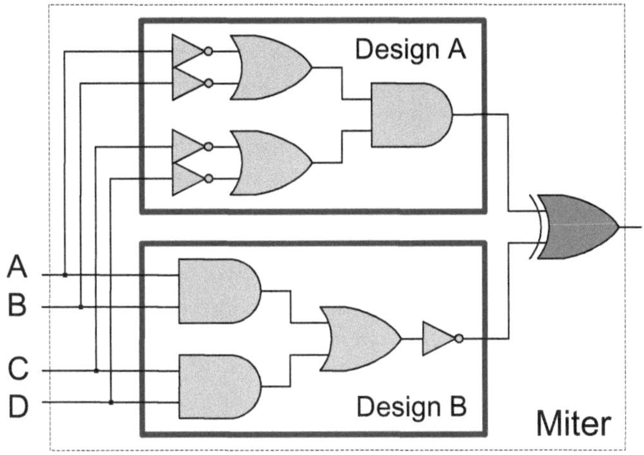

Fig. 2.4 Miter circuit construction for equivalence checking. To check that two versions A and B of a design implement the same combinational logic function, a miter is built by tying corresponding primary inputs together and *xor*-ing corresponding primary outputs. Several techniques, including formal verification engines such as Binary Decision Diagrams and SAT solvers can then invoked to prove that under the outputs of such miter circuit can never evaluate to 1, under all possible input combinations.

2.1.5 Emulation and beyond

As mentioned before, RTL simulation is several orders of magnitude slower than simulation at the architectural level. However, its performance can be improved by leveraging emulation techniques. In *emulation*, also called fast-prototyping, a design is mapped into programmable hardware components, such as FPGAs. These components can be configured after their manufacturing to implement any logic function in hardware and, thus, can be used to create early prototypes of complex systems before they are manufactured. Several companies provide fast prototyping solutions based on FPGAs or other specialized reconfigurable components. It is typical for these systems to have lower performance and lower device density than the final manufactured version of the system under verification. On the other hand, emulation can be conducted on a hardware prototype at much better performance than in a software simulator. Modern FPGAs can run at speeds of hundreds of megahertz (about one order magnitude slower than today's processors) and one may need dozens of them to model a single processor. Note also that, when a block of logic is mapped to an FPGA, its internal signals become invisible to the engineer, thus error diagnosis may become more challenging, unless other means of access to internal signals are specifically implemented in the prototype. Despite these shortcomings, the emulation of a processor design is an important and useful step in pre-silicon verification, since it allows a detailed netlist description to be executed at fairly high performance, enabling testing with longer sequences of stimuli and achieving higher design coverage.

After the design is deemed sufficiently bug-free, that is, satisfactory coverage levels are reached and the design is stable, the device is *taped out*, *i.e.,* sent to the fabrication facility to be manufactured in silicon. Once the first few prototypes are available, the verification transitions into the post-silicon phase, which we describe in the following section. It is important to remember, that the pre- and post-silicon phases of processor validation are not disjoint ventures - much of the *verification collateral*, generated in early design stages, can and should be reused for validation of the manufactured hardware. For instance, random test generators, directed tests and regression suites can be shared between the two. Moreover, in the case of randomized tests, architectural simulators can be used to check the output of the actual hardware prototypes for correctness. Finally, RTL simulation and emulation provide valuable support in silicon debugging. As we explain later, observability of the design's internal signals in post-silicon validation is extremely limited, therefore, it is very difficult to diagnose internal design conditions that manifest in to an error. To alleviate this, test sequences exposing bugs may be replayed in RTL simulation or emulated, to narrow the region affected by the problem and to diagnose the root cause of the error.

In summary, the pre-silicon verification of a modern processor is a complex and arduous task, which requires deep understanding of the design and the capabilities of validation tools, as well as good planning and management skills. After each design transformation, several important questions must be answered: what needs to be verified? how do we verify it? and how do we know when verification is satisfactory? Time is another important concern in this process, since designers must meet stringent schedules, so their products remain competitive in the market. Therefore, they must often forgo guarantees of full correctness for the circuit, and rely on coverage and other indirect measures of validation thoroughness and completeness. Exacerbating the process further is the fact that the design process is not a straightforward one - some modules and verification collaterals may be inherited from previous generations of the product, and some may be only available in a low level description. For instance, performance-critical units may be first created at gate or transistor level (so called *full-custom design*), and the corresponding RTL is generated later. Moreover, often the verification steps discussed in this section must be revisited several times, *e.g.,* when a change in one unit triggers modifications to others components. Thus, pre-silicon verification goes hand in hand with design, and it is often hard to separate them from each other.

2.2 Post-silicon Validation

Post-silicon validation commences when the first prototype of a design becomes available. With the actual silicon in hand, designers can test many physical characteristics of the device, which could not be validated with models of the design. For instance, engineers can determine the *operational region* of the design in terms of such parameters as temperature, voltage and frequency. Physical properties of the device can also be evaluated using detailed transistor-level descriptions, as discussed in the previous section but, when using such models, these analyses can only be conducted on fairly small portions of the design and on very short execution sequences, making it difficult to attain highly quality measures of electrical aspects. For instance, the operational region of the device cannot be evaluated precisely in pre-silicon and must be checked after the device is manufactured. The actual operational region of the design is compared to the requirements imposed by the specification, driven by the market sector targeted by this product. Processors for mobile platforms, for example, usually must operate at lower voltages, to reduce power consumption, while server processors, for which performance is paramount, can tolerate higher temperatures, but also must run at much higher frequency. Other electrical properties that are also typically checked at this stage include drive strength of the I/O pins, power consumption, *etc.*

Most of the electrical defects targeted during post-silicon validation are first discovered as functional errors. For instance, incorrectly sized transistors on the die may result in unanticipated critical paths. Consequently, when the device is running at high frequency, occasionally data will not propagate through these paths within a single cycle, resulting in incorrect computation results. Similarly, jitter in the clock signal may cause internal flip-flops to latch erroneous data values, or cross-talk between buses may corrupt messages in flight. Such bugs are frequently first found as failures of test sequences to which the prototype is subjected. Designers proceed then to investigate the nature of the bug: by executing the same test sequence on a other prototypes they can determine if the bug is of electrical or functional nature. In fact, functional bugs will manifest on all prototypes, while electrical ones will only occur in a fraction of the chips. Bugs that manifest only in a very small portion of the prototypes are deemed to be due to manufacturing defects. Because each of these issues are diagnosed with distinct methods, a correct classification is key to shorten the diagnosis time.

In debugging electrical defects, engineers try to first determine the boundaries of the failure, *i.e.,* discover the range of conditions that trigger the problem, so it can be reproduced and analyzed in the future. To this end *shmoo plots* that depict failure occurrences as a function of frequency and supply voltage are created. Typically, multiple shmoo plots for different temperature settings are created, adding the third dimension to the failure region of the processor. This data can then be analyzed for characteristic patterns of various bug types. For instance, failures at high temperatures are strong indicators of transistor leakage and critical path violations, while errors that occur at low temperatures are often due to race conditions, charge sharing, *etc.* [JG04]. Designers then try to pinpoint the area of the circuit where the error occurs by adjusting the operating parameters of individual sub-modules with techniques such as on-die clock shrinks, clock skewing circuits [JPG01] and optical probing , which relies on lasers to measure voltage across individual transistors on the die [EWR00]. In optical probing, the silicon substrate on the back side of the die is etched and infrared light is pulsed at a precise point of the die. The silicon substrate is partially transparent to this wavelength, while doped regions of transistors reflect the laser back. If electrical charge is present in these regions, the power of the reflected light changes, allowing the engineers to measure the voltage. Note that in this case, the top side of the processor, where multiple metal layers reside, does not need to be physically etched or probed, so integrity of the die is not violated. Unfortunately, the laser cannot get to the back side of the die through a heat sink, which, therefore, must be removed around the sampling point. While providing good spacial and timing resolution, optical probing remains a very expensive and often ineffective way of testing, since it requires sophisticated apparatus and enables access to only a single location at a time. Finally, when the issue is narrowed down to a small block, transistor-level simulation can be leveraged to establish the root cause of the bug and determine ways to remedy it.

As we mentioned above, in addition to electrical bugs, two types of issues can be discovered in manufactured prototypes: fabrication defects and functional errors, which are the target of *structural testing* and *functional post-silicon validation*, respectively. Although similar at first glance, these approaches have a very important difference in that testing assumes that the pre-silicon netlist is functionally correct and tries to establish if each prototype faithfully and fully implements it. Validation, on the other hand checks if the prototype's functionality adheres to the specification, that is, if the processor can properly execute software and correctly interact with other components of a computer system. In the following section we overview structural testing approaches and discuss silicon state acquisition techniques, which are typically deployed in complex designs to improve testability. Incidentally, most of these acquisition solutions can be also used in post-silicon validation, discussed in Section 2.2.2.

2.2.1 Structural testing

Modern integrated circuits are manufactured with a photolithographic process, which "prints" individual transistors, as well as metal interconnect between them, in multiple layers onto a silicon substrate. However, due to the nanometer-scale of transistors and wires, their features' boundaries may be blurry or distorted. This may cause, for example, two metal paths to be shorted together, or result in misalignment of the doped regions of a transistor, impairing the overall functionality of the circuit. The phase of post-silicon validation that tries to uncover these faults is called *structural testing*. In this framework, the gate-level netlist of the design is assumed as the golden model of functionality and used by *automatic test pattern generators* (ATPGs) in producing test sequences that check these design aspects. An ATPG analyzes the combinational logic of the netlist and infers test vectors that would expose a range of possible defects generated in the manufacturing process. In a sense, ATPG-based structural testing can be thought of as equivalence checking between the pre-production and the printed netlists. The type of faults that ATPGs can discover include shorts and opens, "stuck-at" defects, where a wire's value never changes, and violations of the circuit's internal propagation delays. ATPG patterns are then applied to the silicon prototype and the responses are compared to those predicted by simulation on the pre-production netlist. For example, to test an implementation of a two-input logical *and* gate, an ATPG solution would check that the output of the silicon prototype is consistent with the truth table of the function, that is only when both inputs are high, the output value is equal to one. Modern testing tools typically do not subject the design to an exhaustive set of test vectors, which may be prohibitively large, but use advanced heuristics to minimize the set of test patterns without loss of defect coverage. Note, however, that ATPG testers cannot discover functional errors in the circuit, that is, discrepancies between design intent and the implemented silicon prototype.

In addition to its inability to discover functional errors in the circuit, the scalability of structural testing is severely limited in sequential designs: *i.e.,* systems that have internal storage elements in addition to combinational logic blocks. This is especially pronounced in complex microprocessor designs, where data is retained by internal storage elements for many cycles. As a consequence, it is virtually impossible for an ATPG technique to create test vectors to be applied to primary inputs of the circuit that can test behavior of logic functions deep inside of the design and control all types of manufacturing faults. Likewise, it may be impossible to propagate the information about the error to primary outputs to be observed by the designer. Faced with the dual problem of *controllability* and *observability* in such complex sequential circuits, it is mainstream today to augment the design with structures that allow comparatively easier access to internal logic nodes through I/O pins. This is commonly referred to as *design for testability*, or DFT. In particular, DFT techniques often provide ways to sample and write state elements of the circuits, such as flip-flops and latches, so combinational logic can efficiently be tested by ATPG patterns. Furthermore, as discussed later in this chapter, DFT techniques play an important role in functional post-silicon validation and debug, where they are used to analyze the internal behavior of a prototype that leads to an error. Research on structural testing and DFT has been carried out for decades in both industry and academia and it would be impossible to overview the most successful techniques within the scope of this section. Therefore, we will limit ourselves to briefly discuss a handful of the most notable solutions for silicon state acquisition, and recommend two textbooks as starting points for a more in depth study: "Digital Systems Testing and Testable Design" [ABF94] and "Essentials in Electronic Testing" [BA00].

One of the most basic and classic techniques in the DFT domain are *scan chain* components, an example of which is shown in Figure 2.5. To include a scan chain in a design, flip-flops are augmented with an additional data input (*scan in*) and output (*scan out*), as well as an enable line. The scan in and scan out lines of different storage elements are then connected serially into a *scan chain*, the ends of which are tied to special I/O pins of the device. When the scan enable signal is de-asserted, the flip-flops act as regular storage cells and the processor operates in normal mode. When enable is asserted, on the other hand, the scan chain reconfigures itself into a serial register, so that data can be passed serially from one flip-flop to another. With this tool at hand, verification engineers can suspend execution and scan out the values of the chained storage elements through a single output wire and analyze it. Moreover, an arbitrary internal state can be pre-set through the scan in functionality, enabling

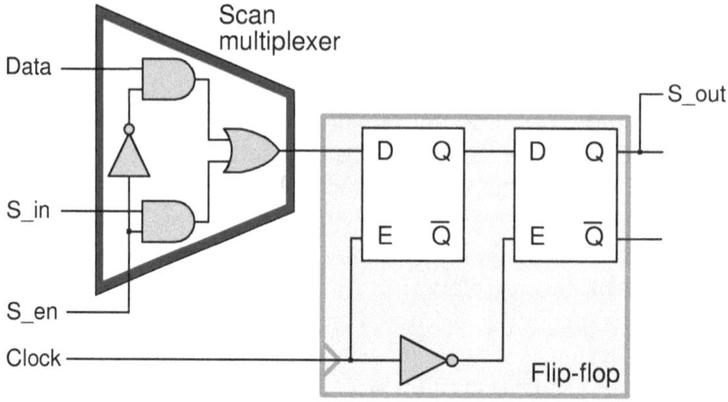

Fig. 2.5 Scan flip-flop design. To insert a regular D flip-flop in a scan chain, the flip-flop is augmented with a scan multiplexer, which selects the source of data to be stored. During regular operation, the S_en (scan enable) signal is de-asserted and the flip-flop stores bits from the Data line. When enable is asserted, however, the flip-flop samples the scan in (S_in) signal instead. This allows the designers to create a scan chain, by connecting the scan out (S_out) output with a scan in input of the next flip-flop in the chain. The last output in this chain is connected to a dedicated circuit output port, so that the internal state of the system may be shifted out by asserting S_en signal and pulsing the clock. Likewise, since the S_in input of the first chain element is driven from a circuit's primary input, engineers can quickly pre-set an arbitrary internal state in the design for testing and debugging purposes. Note that during scan chain operations the regular design functionality is suspended.

fine-grain controllability of the device, in addition to observability. Particular implementations of the scan technique vary in different designs and include full-scan (all state elements are connected), partial scan (a subset of flip-flops is connected), and multiple parallel chains. The drawbacks of the approach are a somewhat slower latch operation, the additional area and interconnect overheads, and, most importantly, the need to suspend operation of the device under test to shift the state out or bring a new state in, an activity that requires several clock cycles.

Modern scan chains often rely on an even more complex scan flip-flop design, that overcomes the limitation of having to suspend execution while routing state in and out of the system. *Hold-scan* flip-flops consists of two basic scan flip-flops connected together, called a primary and a shadow flip-flop, as shown in Figure 2.6. When the capture line is asserted, the shadow element samples and stores the value of the main latch. This flip-flop can be reconfigured into a scan chain connection at any later time, so the sampled data can be routed outside of the device. Since the chain comprises only shadow flip-flops and operates on an independent clock, the device can continue to function normally, while the captured snapshot is being transferred out. Similarly, a new state can be loaded into the shadow latches without interrupting the operation of the device and then propagated to main flip-flops by the asserting the update line. Although hold-scan flip-flops have significantly larger area than the baseline design of Figure 2.5, it provides the designers with much more flexibility in the debugging process and, consequently, and it is frequently deployed in microprocessor systems [KDF$^+$04].

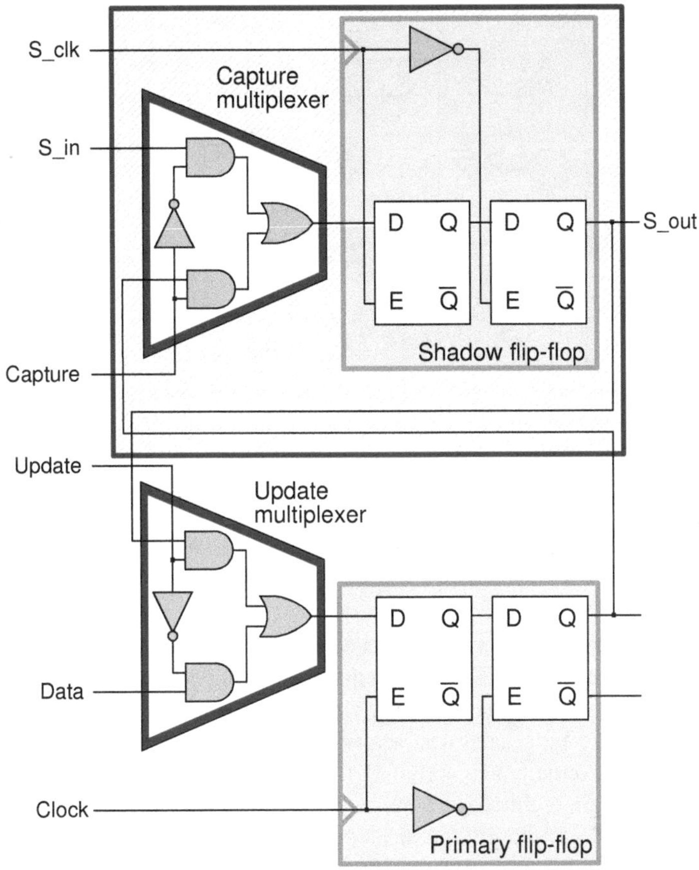

Fig. 2.6 Hold-scan flip-flop design. Hold-scan flip-flops provide the ability to overlap system execution with scan activity. The component comprises two scan flip-flops, a primary and a shadow flip-flop. The shadow flip-flop can capture and hold the state of the primary storage element. Shadow elements are connected in chain and can transmit the captured values without the need to suspend regular system operation. Similarly they can be used to load a system state, which is then transferred to the primary flip-flops by asserting the update signal. Note that shadow flip-flops operate on a separate clock (S_clk), so that the transmit frequency can be decoupled from the system's operating frequency.

Boundary-scan is another technique often used in structural testing and validation, to allow individual modules of the processor to be tested in isolation [IEE01a]. Boundary-scan was developed by the Joint Test Action Group (JTAG), which was formed by the industry to enable testing of complex circuits boards. JTAG calls for inputs and outputs of the chip, or a block of logic, to be tied to dedicated, scan storage elements, as shown in Figure 2.7. These cells can be configured to perform several distinct functions by means of specialized control logic. For instance, inputs to the block can be captured to verify operation of other modules or can be used provide stimulus to the block. The response to the stimulus can be then observed

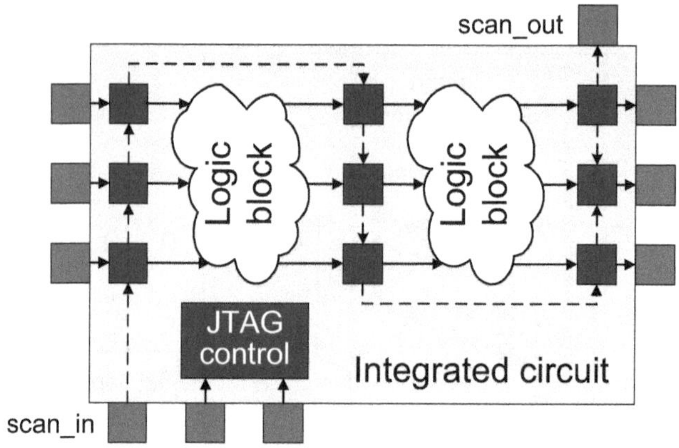

Fig. 2.7 Boundary-scan technique (JTAG). Scan storage cells are inserted and connected to the I/O pins of a chip or at the boundary of internal logic blocks. These cells are also chained together as indicated by the dashed lines in the figure. Through an additional on-die JTAG controller, designers can control the operation of the cells and use them to test the interconnect among chips on the same board or among blocks within a chip, and test individual blocks in isolation.

in the boundary-scan flip-flops tied to its outputs. Today, JTAG technology is often extended to perform additional configuration and testing operations, such as on-chip programming, access to special memory or registers, *etc.*

Despite all these functionalities, scan-based techniques still lack an important feature, that is, automatic acquisition of the internal state triggered by a design's behavior. Scan chain technology requires that an enable pin is externally asserted in order to initiate sampling. On-chip logic analyzers (OCLA) alleviate this limitation, augmenting the device with programmable trigger-detection and response logic, thus making the sampling activity programmable [CKM$^+$04]. OCLAs are designed to observe internal signals continuously and to match them against preset triggering conditions such as, for instance, a signal assuming a specific, user-programmed logic value or undergoing a certain transition. When a triggering condition occurs, the analyzer begins a continuous sampling of the design's state for a given time period and stores the data into dedicated memory. A designer can later download this data and monitor the behavior of the circuit throughout several cycles of execution. Unfortunately, the advanced functionality of on-chip logic analyzers comes at the price of high area penalty and the need to route the sampling wires throughout the die. In the highly competitive settings of the microprocessor industry, this area overhead can rarely be afforded, compelling companies to limit the use of OCLAs [Lit02] and/or diagnose bugs by relying simply on the scan-based testing and verification techniques overviewed above.

2.2.2 Functional post-silicon validation

Identification of functional bugs is another crucial part of the processor design and manufacturing effort, whose importance has been quickly growing in recent years. As we mentioned earlier, functional post-silicon validation differs from structural testing in that it does not target errors due to manufacturing issues, but strives to establish if the prototype adheres to its specification and faithfully embodies the designer's intent. The main advantage that designers can leverage in post-silicon, compared to pre-manufacturing validation, is the test execution speed. Indeed, instead of an architectural or RTL simulation, in post-silicon validation programs execute on the actual hardware, outperforming the former two techniques by multiple orders of magnitude. Therefore, significantly longer sequences can be tested and, consequently, higher coverage may be attained. To put things in perspective, consider that during the many months of validation of the Pentium 4 processor, an experience discussed in Section 1.3, a total of about 200 billion cycles were simulated before tape-out [Ben01]. After manufacturing, these accounts for only three minutes of runtime in the actual processor, which operates at 1GHz frequency. Functional post-silicon validation is carried out through directed and parameterized random tests. The drawback of the latter is the lack of a known correct outcome, needed to determine the correctness of the silicon prototype's output. Thus, to find the correct responses, engineers turn to architectural simulators. These have much lower performance than prototype hardware, and thus vendors are often forced to employ large server farms to simulate the same random tests that they run on silicon and derive correct outputs to compare against the prototype's results [Rot00]. Nevertheless, verification with randomized programs remains a central component of the post-silicon validation process, since it often exposes many unexpected behaviors of the system missed by directed tests. Following the same rationale, designers often randomize the manner in which primary inputs of the prototype, such as external interrupts, are asserted during a test and alter the delay of messages sent to and from the processor on the system bus. Typically, this is done with custom hardware components, which reside on the same board as the prototype under test and can be programmed to exhibit a variety of behaviors [SFH+03].

In addition to the functional validation of the prototype itself, the post-silicon phase must also evaluate the compatibility of the processor with a number of deployment platforms. For this purpose, the device is plugged into a typical system board with commercially available peripherals. On this complete system the verification team runs a broad range of directed benchmarks and software applications. Examples include the boot up software of several operating systems, commercial applications, performance benchmarks, legacy software, and so on. The operating conditions of the prototype in this case, are not as stressful as with randomized tests, however, these tests are still necessary to establish system compatibility throughout hours of actual runtime.

Identifying an erroneous behavior in functional post-silicon validation is only the beginning of a long and arduous journey to determine the root cause of the problem and fix it. As with electrical defects, the process begins with trying to reproduce the bug and determine the conditions under which it occurs. Throughout this activity, DFT techniques, discussed above, play a vital role, since with these solutions designers can sample the internal state of the prototype and unwind the events that caused erroneous output behavior all the way back to the bug. In addition to generic state acquisition solutions, such as scan and OCLA, in the microprocessor industry, engineers often deploy a number of domain-specific techniques, which take advantage of the system's built-in performance counters and special interrupt modes. For instance, an external interrupt can be programmed to invoke a software routine which dumps the processor's state to memory, and then it can be asserted several times during a test execution to facilitate bug diagnosis [SFH$^+$03]. It is important to note, however, that silicon state acquisition techniques such as this may alter the timing of events occurring within the processor, obfuscating the error.

Once a set of traces leading to an error is obtained, engineers can go back to pre-silicon techniques to reproduce and identify the bug in simulation or through formal methods. Since pre-silicon simulation has much lower performance than a prototype, full replay of long traces – typical of post-silicon runs – cannot be completed in reasonable time. Two possible remedies to this problem are trace minimization, a technique that can shorten a trace while still exposing the bug under study, *e.g.*, [CBM07b, SVM07], or the identification of a subset of events necessary for the occurrence of the error, typically accomplished by state analysis through DFT structures. A shorter trace can then be replayed in simulation, perhaps at full chip-level first and then at block granularity, and the root cause of the issue can be established. At this point formal tools may also be used to identify conditions under which the error manifest, as well as error location in an RTL design description [CWBM07]. The intermediate states that the design enters on its path from the reset condition to the bug can be used for guidance, in practice decomposing the problem into smaller subproblems, each solvable formally [HTB$^+$09]. In addition to pinpointing the error location, formal analysis can be used to suggest the way the issue can be corrected to restore proper design functionality [SV06, CBM07a]. Finally, when the issue is diagnosed, designers propose corrections that are evaluated in pre-silicon first and then put into new versions of the prototype. Because silicon manufacturing is a costly and time consuming process, engineers always strive to develop work-arounds for a problem so to continue the validation of a prototype as long as possible without the need for a *re-spin*, that is a new tape-out. Moreover, non-critical bugs not get fixed through hardware design modifications, and instead may be simply documented with the processor's *errata* or be overcome through other non-hardware solutions.

Despite a wide range of solutions available in this domain, post-silicon validation remains a difficult and expensive activity. Processor vendors are forced to invest into such expensive hardware as high-speed logic analyzers, optical probing machines and simulation server farms, in addition to employing large teams of professional verification engineers. As a consequence, design teams have traditionally relied on high-quality pre-silicon verification to expose the majority of errors in a processor. Unfortunately, with the growing complexity of today's processors and lagging speeds of RTL simulation and verification, this is becoming harder and harder to accomplish. Novel technologies, such as hyper-threading, multi-core architectures, variable-delay on-die interconnect and virtualization, exacerbate the problem further, stretching to the limit pre-silicon verification capabilities of even the largest processor vendors. Post-silicon functional verification has been called in to pick up the slack, and as a consequence, the fraction of design costs devoted to post-silicon validation has been increasing steadily in recent years [Yer06]. In the future, we expect this trend to continue, and anticipate the appearance of novel and exciting processor features that advance the capabilities of post-silicon validation.

2.3 Runtime Verification

In the previous sections we covered the main pre- and post-silicon validation steps, occurring behind the doors of the design house, before a processor is released to the market. With these techniques, the engineering team detects the vast majority of design bugs. However, a very small fraction may, and often does, escape into the final silicon product. The reasons for this lack of completeness lie in the inherent limitation of pre- and post-silicon technology: formal pre-silicon approaches are not scalable to designs as large as today's microprocessors and require that device's specifications and verification goals are carefully crafted as mathematical expressions - a feat unachievable in many cases. Simulation-based pre-silicon techniques, on the other hand, can be applied to significantly larger designs, but, in return, they provide much lower levels of correctness guarantees. In other words, simulation can only vouch for the error-free execution of cases that were exercised during the test execution and, consequently, bugs can be left lurking in the areas of a design unexplored by the tests. Furthermore, as the design is taken from its architectural description to RTL, the level of detail increases, reducing simulation performance by orders of magnitude. As a result, only fairly short test sequences can be executed on a full-system RTL model, before the design is manufactured. Post-silicon validation exposes a silicon prototype to much harsher and longer testing, revealing additional bugs that could not be discovered before manufacturing, such as electrical and manufacturing defects, as well as hard-to-reach functional bugs. The crucial shortcoming of this development phase is the lack of observability of the device's internal signals and state. While silicon state acquisition methods alleviate this process, hardware prototype diagnosis and debugging remains to this day one of the most challenging design tasks.

Overall, both pre- and post-silicon validation can expose the vast majority of bugs in a modern microprocessor, but, unfortunately, they cannot identify all of them in a reasonable time. Thus, there is always a small number of *escaped bugs*, which find their way into the final silicon product shipped to end-users. Sometimes vendor companies make information about such errors available, publishing specification updates or *errata sheets*. Typically, these escaped bugs occur rarely – which is why they are very hard to find – and may include cases of unanticipated interactions between on-chip components or between a processor and the peripherals connected to it. Nevertheless, they can cause significant damage to the company, which may be forced to issue a recall or delay the release of the product. More importantly, unlike buggy software, circuits printed onto a silicon substrate cannot be easily modified to correct an issue, especially if they reside in millions of computers distributed all over the world. Therefore, processor vendors and researchers in academia alike have began developing solutions to guarantee correct operation of their products in the field, even in presence of hardware bugs, with so-called *dynamic*, or *runtime verification* techniques.

The most primitive form of runtime verification consists of modifying any software application running on the hardware platform to avoid the occurrence of the erroneous sequence. This method is referred to as *instruction patching* and it is most commonly used in the embedded domain, where both hardware and the lower-level software, such as drivers and system libraries, are strictly controlled by the same vendor. In general purpose computing platforms, however, it is practically impossible to rewrite all possible software applications that a buggy processor may execute once in the field. In these platforms, engineers attempt to disable the functional blocks responsible for the issue, whenever possible. This can be accomplished through basic input-output system (BIOS) re-configuration. A BIOS is bootable firmware program stored in non-violatile memory on the motherboard of the system, the program executes at computer startup to configure the system before an operating system is loaded. To allow in-the-field system reconfiguration (through component disabling), designers augment some blocks of the processor with special programmable flags, sometimes known as *chicken bits*. If a bug is discovered that can be isolated to one of this logic blocks, the manufacturer can re-write the BIOS firmware so to disable it at startup, and thus fix the issue. However, disabling a hardware module will often lead to lower system performance or, possibly, a reduced feature set. For instance, in a recent AMD's Phenom processor a look-aside buffer had to be disabled in the cache to avoid a race condition that would hang the whole system. The consequence was a 10% performance drop [AMD08], an amount sufficient to cause a notable financial impact to the manufacturer. An alternative solution that exercises more fine-grain control on the design's features, and, therefore, has lower performance penalty, was developed at Intel after the FDIV error fiasco [SC04]. This technique, called *microcode update*, allows the processor to change the semantics of execution of individual ISA instructions. Essentially, the decode stage of a processor pipeline is made programmable and, through firmware updates, it becomes possible to re-program the execution semantics of individual instructions. At runtime, when instructions leading to a bug are fetched and decoded, the execution sequence for those instruction is replaced based on the firmware information and the occurrence of the bug can be circumvented. Note that, similar to BIOS firmware update, microcode updates must be developed offline by the processor vendor company after the escaped error is discovered in a released product and the root cause of the bug is diagnosed.

While microcode update technology faithfully served the industry in the last decade, it was not without problems of its own. For instance, AMD's implementation of this technology itself was found at some point to contain an uncorrectable bug [AMD03]. More importantly, as the complexity of processors increased and multi-core designs appeared on the consumer market, this relatively simple technology became insufficient to detect and patch errors arising from complex interactions between on-die modules. Subsequently, researchers in academia have began investigating approaches that enable error patching at the level of individual control logic signals [WBA06, SNC+07]. These techniques augment the processor with programmable pattern matchers, which monitor control logic signals of the device and compare them with bug patterns loaded at startup. When a match occurs, an error recovery mechanism is triggered to bypass the issue. Patterns describing the bugs, also called bug signatures or fingerprints, are created and distributed by processor vendors similarly to BIOS and microcode updates.

The most important drawback of all patching-based approaches, however, is the fact that a significant amount of time passes between when a bug is first discovered and reported and when a patch is released. During this time the error could be exploited by malicious users to breach the security of a computer system or render it inoperable. Therefore, in many applications it is desirable to be able to guarantee the correctness and security of a processor's operation even in presence of unknown escaped errors. Traditionally, in domains where errors cannot be tolerated, such as aero-space, or military applications, this goal has been accomplished through N-modular redundancy design, where multiple distinct implementations of the same hardware are deployed together and operate concurrently. All modules in this case are derived from a same specification, but are implemented independently and, therefore, are unlikely to contain identical errors. At runtime, the modules operate concurrently and errors are detected (and possibly corrected) when results differ among the various implementations. Although this redundancy may provide fairly good protection from errors, its cost and area overheads are far beyond what can be afforded in commercial processors. As a result, in recent years, researchers have proposed a number of novel, low cost, runtime verification solutions. Many of these solutions are based on hardware checkers, small blocks encoding invariants of correctness for the design's operation: for instance they can be used to dynamically validate the operation of individual cores [Aus00, DBB07] or the communication between them [MS05]. Other solutions have also proposed the insertion of local checkers distributed throughout the design to monitor the correctness of functional invariants [NdPC+04, BM05]. When an error is detected, a recovery mechanism is invoked, so the system can circumvent the troublesome spot and then return to normal execution. The key difference between checker- and patching-based solutions is that the former encode the correct scenarios of processor operation, while the latter compare the state of the design with known erroneous patterns. Typically, checker-based techniques incur higher area penalty; on the other hand, they ensure operation of the system even in the presence of undiscovered bugs.

Overall, runtime verification is a new and exciting field in processor design and validation, which promises to dramatically improve the quality of commercial processors and protect both vendors and end-users from escaped errors. Solutions in this area have just began to appear in the research literature and are yet to be fully embraced by the industry. However, we believe that the growing complexity of modern processors will inevitably lead to the deployment of runtime verification techniques into commercial designs. Nevertheless, they will not become a substitute for pre- and post-silicon validation efforts. Each one of these domains has unique and irreplaceable advantages, and it is most likely that all of them will come together into a balanced solution, complementing each other and providing utmost efficiency in microprocessor validation.

2.4 Summary

In this chapter we presented a broad overview of verification techniques at all levels of a processor's life-cycle: from pre-silicon to in-the-field deployment. As we reported, verification goes hand in hand with the design process, and often must quickly adapt to modification in the specifications of a device. In pre-silicon verification, which is conducted on a range of software models of the design, simulation-based techniques remain the workhorse of the the industry, primarily, due to their scalability. On the other hand, formal verification, where a processor is subjected to rigorous mathematical proofs, does not scale to large designs, but can provide significantly higher guarantees of correctness. Therefore, formal tools are often used to selectively target relatively small internal modules of a processor that are more prone to errors and are of vital importance to the functionality of the device.

Post-silicon validation commences once a prototype of the processor is manufactured and targets electrical, functional and manufacturing defects. The latter ones are often diagnosed with efficient ATPG solutions, which subject the prototype to a variety of structural test vectors. To achieve efficient testing and debugging in this domain, engineers gain access to the internal state of a prototype through a range of DFT mechanisms, such as scan-chains, JTAG, *etc.* Functional post-silicon validation, on the other hand, aims to establish that a design truly adheres to its original specification and operates properly in a wide range of computing platforms. In particular, functional post-silicon validation in today's industry consists of executing a mix of directed and randomized tests on the hardware prototype, and then validating the corresponding outcomes against those produced by an architectural simulator or checking for a pre-designed result.

When the prototype is deemed sufficiently stable in the post-silicon phase, it is released to the market and deployed in the field. Due to short production schedules in today's competitive market, which often result in insufficient pre- and post-silicon validation, processors are frequently released with a small number of subtle bugs. To bypass such errors in the field engineers make their designs re-configurable through BIOS and firmware patches, or augment them with hardware checkers to monitor execution correctness, in addition to providing recovery mechanisms upon bug detection. Each of the three classes of verification solutions, pre-silicon, post-silicon and runtime, has unique advantages and drawbacks and therefore, all of them must used in concert to achieve the highest level of protection from errors and to ensure minimum performance, cost and silicon area impact.

References

[AAF+04] Allon Adir, Eli Almog, Laurent Fournier, Eitan Marcus, Michal Rimon, Michael Vinov, and Avi Ziv. Genesys-pro: Innovations in test program generation for functional processor verification. *IEEE Design & Test of Computers*, 21(2):84–93, 2004.

[ABF94] Miron Abramovici, Melvin A. Breuer, and Arthur D. Friedman. *Digital Systems Testing and Testable Design*. Wiley-IEEE Press, revised edition, September 1994.

[Acc04] Accellera. *Property Specification Language Reference Manual, Rev 1.1*, June 2004. `http://www.eda.org/vfv/docs/PSL-v1.1.pdf`.

[AMD03] Advanced Micro Devices, Inc. *AMD Athlon™ Processor Model 6 Revision Guide*, October 2003. `http://www.amd.com/us-en/assets/content_type/white_papers_and_tech_docs/24332.pdf`.

[AMD08] Advanced Micro Devices, Inc. *Revision Guide for AMD Family 10h Processors*, April 2008. `http://.amd.com/us-en/assets/content_type/white_papers_and_tech_docs/41322.PDF`.

[AMS08] Charles J. Alpert, Dinesh P. Mehta, and Sachin S. Sapatnekar. *Handbook of Algorithms for Physical Automation*. Taylor and Francis, first edition, 2008.

[Arv03] Arvind. Bluespec: A language for hardware design, simulation, synthesis and verification. In *MEMOCODE, Proceedings of the ACM/EEE International Conference on Formal Methods and Models for Co-Design*, page 249, June 2003.

[Aus00] Todd Austin. DIVA: A dynamic approach to microprocessor verification. *Journal of Instruction-Level Parallelism*, 2:1–26, May 2000.

[BA00] Michael Bushnell and Vishwanti Agrawal. *Essentials of Electronic Testing for Digital, Memory and Mixed-signal VLSI Circuits*. Springer, 2000.

[BA08] David Burger and Todd Austin. The SimpleScalar toolset, version 3.0, 2008. `http://simplescalar.com`.

[BAWR07] Jayanta Bhadra, Magdy S. Abadir, Li-C. Wang, and Sandip Ray. A survey of hybrid techniques for functional verification. *IEEE Design and Test of Computers*, 24(2):112–122, March 2007.

[BBB+87] Randal E. Bryant, Derek L. Beatty, Karl Brace, Kyeongsoon Cho, and Thomas Sheffler. COSMOS: A compiled simulator for MOS circuits. In *DAC, Proceedings of the Design Automation Conference*, pages 9–16, June 1987.

[BBS90] Derek L. Beatty, Randal E. Bryant, and Carl-Johann H. Seger. Synchronous circuit verification by symbolic simulation: An illustration. In *Proceedings of Sixth MIT conference on advanced research in VLSI*, pages 98–112, April 1990.

[Ben01] Bob Bentley. Validating the Intel® Pentium® 4 microprocessor. In *DAC, Proceedings of the Design Automation Conference*, pages 224–228, June 2001.

[Ber05] Valeria Bertacco. *Formal Verification Scalable Hardware Verification With Symbolic Simulation*. Springer, first edition, 2005.

[BLL+04] Michael Behm, John Ludden, Yossi Lichtenstein, Michal Rimon, and Michael Vinov. Industrial experience with test generation languages for processor verification. *DAC, Proceedings of the Design Automation Conference*, pages 36–40, June 2004.

[BM05] Ali A. Bayazit and Sharad Malik. Complementary use of runtime validation and model checking. In *ICCAD, Proceedings of the International Conference on Computer Aided Design*, pages 1052–1059, 2005.

[Boc07] Bochs: The open source IA-32 emulation project, September 2007. `http://bochs.sourceforge.net/`.

[BRB90] Karl Brace, Richard Rudell, and Randal E. Bryant. Efficient implementation of a BDD package. In *DAC, Proceedings of the Design Automation Conference*, pages 40–45, June 1990.

[Bry85] Randal E. Bryant. Symbolic verification of MOS circuits. In *Proceedings of Chapel Hill Conference on VLSI*, pages 419–438, May 1985.

[Bry86] Randal E. Bryant. Graph-based algorithms for Boolean function manipulation. *IEEE Transactions on Computers*, 35:677–691, 1986.

[CBM89] Oliver Coudert, Christian Berthet, and Jean C. Madre. Verification of synchronous sequential machines based on symbolic execution. In *Proceedings of the International Workshop on Automatic Verification Methods for Finite State Systems*, pages 365–373, June 1989.

[CBM07a] Kai-hui Chang, Valeria Bertacco, and Igor Markov. Automating post-silicon debugging and repair. In *ICCAD, Proceedings of the International Conference on Computer Aided Design*, pages 91–98, November 2007.

[CBM07b] Kai-hui Chang, Valeria Bertacco, and Igor Markov. Simulation-based bug trace minimization with BMC-based refinement. *IEEE Transactions on Computer-Aided Design of Integrated Circuits and Systems*, 26(1):152–165, 2007.

[CBRZ01] Edmund M. Clarke, Armin Biere, Richard Raimi, and Yunshan Zhu. Bounded model checking using satisfiability solving. *Formal Methods in System Design*, 19(1):7–34, 2001.

[CGP08] Edmund M. Clarke, Orna Grumberg, and Doron A. Peled. *Model Checking*. MIT Press, sixth edition, 2008.

[CJB79] William C. Carter, William H. Joyner, and Daniel Brand. Symbolic simulation for correct machine design. In *DAC, Proceedings of the Design Automation Conference*, pages 280–286, June 1979.

[CKM$^+$04] William D. Corti, Robert Kenny, Joseph O. Marsh, Steven C. Parker, Frank X. Scanzano, and Michael Won. *U.S. Patent no. 6834360: On-chip logic analyzer*. International Business Machines Corporation, December 2004.

[CVK04] Ben Cohen, Srinivasan Venkataramanan, and Ajeetha Kumari. *Using PSL/Sugar for Formal and Dynamic Verification*. VhdlCohen Publishing, second edition, 2004.

[CWBM07] Kai-hui Chang, Ilya Wagner, Valeria Bertacco, and Igor Markov. Automatic error diagnosis and correction for RTL designs. In *HLDVT, Proceedings of the International Workshop on High Level Design Validation and Test*, pages 65–72, November 2007.

[DBB07] Andrew DeOrio, Adam Bauserman, and Valeria Bertacco. Chico: An on-chip hardware checker for pipeline control logic. In *MTV, Proceedings of the International Workshop on Microprocessor Test and Verification*, pages 91–97, December 2007.

[DP60] Martin Davis and Hilary Putnam. A computing procedure for quantification theory. *Journal of the Association for Computing Machinery*, 7(3):201–215, July 1960.

[ES03] Niklas Een and Niklas Sorensson. An extensible SAT-solver. In *Theory and Applications of Satisfiability Testing*, volume 2919 of *Lecture Notes in Computer Science*, pages 502–518. Springer, 2003.

[EWR00] Travis Eiles, Gary Woods, and Valluri Rao. Optical probing of flip-chip-packaged microprocessors. In *ISSCC, Proceedings of the International Solid State Circuits Conference*, pages 220–221, February 2000.

[FKL04] Harry Foster, Adam Krolnik, and David Lacey. *Assertion-based design*. Springer, second edition, 2004.

[FZ03] Shai Fine and Avi Ziv. Coverage directed test generation for functional verification using Bayesian networks. In *DAC, Proceedings of the Design Automation Conference*, pages 286–281, June 2003.

[Glu06] Alon Gluska. Practical methods in coverage-oriented verification of the Merom microprocessor. In *DAC, Proceedings of the Design Automation Conference*, pages 332–337, July 2006.

[GTKS05] Mehmet H. Gunes, Mitchell Thornton, Fatih Kocan, and Stephen Szygenda. A and comparison of digital logic simulators. In *MWSCAS, Proceedings of the Midwest Symposium on Circuits and Systems*, pages 744–749, August 2005.

[HKM01] Faisal I. Haque, Khizar A. Khan, and Jonathan Michelson. *The Art of Verification with Vera*. Verification Central, first edition, 2001.

[HMN01] Yoav Hollander, Matthew Morley, and Amos Noy. The *e* language: A fresh separation of concerns. In *TOOLS, Proceedings of the, International Conference on Technology of Object-Oriented Languages*, pages 41–50, March 2001.

[HS06] Gary D. Hachtel and Fabio Somenzi. *Logic Synthesis and Verification Algorithms*. Springer, first edition, 2006.

[HTB+09] Richard Ho, Michael Theobald, Brannon Batson, J.P. Grossman, Stanley C. Wang, Joseph Gagliardo, Martin M. Deneroff, Ron O. Dror, and David E. Shaw. Post-silicon debug using formal verification waypoints. In *DVCon, Proceedings of the Design and Verification Conference and Exhibition*, pages 1–7, February 2009.

[IEE01a] Institute of Electrical and Electronics Engineers. *Standard test access port and boundary-scan architecture. IEEE Std. 1149.1-2001.*, 2001. `http://ieeexplore.ieee.org/servlet/opac?punumber=7481`.

[IEE01b] Institute of Electrical and Electronics Engineers. *Standard Verilog hardware description language. Std 1364-2001.*, 2001. `http://ieeexplore.ieee.org/servlet/opac?punumber=7578`.

[IEE04] Institute of Electrical and Electronics Engineers. *Behavioural languages - Part 1-1: VHDL language reference manual. IEC 61691-1-1 First edition 2004-10; IEEE 1076.*, 2004. `http://ieeexplore.ieee.org/servlet/opac?punumber=9649`.

[IEE07] Institute of Electrical and Electronics Engineers. *Standard for SystemVerilog - Unified Hardware Design, Specification, and Verification Language. Std 1800-2007*, 2007. `http://ieeexplore.ieee.org/servlet/opac?punumber=4410438`.

[JG04] Doug Josephson and Bob Gottlieb. The crazy mixed up world of silicon debug. In *CICC, Proceedings of the IEEE Custom Integrated Circuits Conference*, pages 665–670, October 2004.

[JPG01] Doug Josephson, Steve Poehhnan, and Vincent Govan. Debug methodology for the McKinley processor. In *ITC, Proceedings of the International Test Conference*, pages 451–460, October 2001.

[KDF+04] Ravishankar Kuppuswamy, Peter DesRosier, Derek Feltham, Rehan Sheikh, and Paul Thadikaran. Full hold-scan systems in microprocessors: Cost/benefit analysis. *Intel Technology Journal*, 08:63–72, February 2004.

[KG99] Christoph Kern and Mark R. Greenstreet. Formal verification in hardware design: a survey. *ACM Transactions on Design Automation of Electronic Systems*, 4(2):123–193, 1999.

[KK94] Ramayya Kumar and Thomas Kropf. *Theorem provers in circuit design: theory, practice, and experience*. Springer Berlin/ Heidelberg, first edition, 1994.

[Lam02] Leslie Lamport. *Specifying systems: the TLA+ language and tools for hardware and software engineers*. Addison-Wesley Professional, 2002.

[Lit02] Timothe Litt. Support for debugging in the Alpha 21364 microprocessor. In *ITC, Proceedings of the International Test Conference*, pages 584–589, October 2002.

[MCE+02] Peter S. Magnusson, Magnus Christensson, Jesper Eskilson, Daniel Forsgren, Gustav Hallberg, Johan Hogberg, Frederik Larsson, Andreas Moestedt, and Bengt Werner. Simics: A full system simulation platform. *IEEE Computer*, 35(2):50–58, 2002.

[Mei93] Gerd Meister. A survey on parallel logic simulation. Technical report, University of Saarland, Department of Computer Science, 1993.

[Men08] Mentor Graphics®Inc. *ModelSim - a comprehensive simulation and debug environment for complex ASIC and FPGA designs*, 2008. `http://www.model.com/`.

[Mic03] Alexander Miczo. *Digital logic testing and simulation*. John Wiley and Sons, second edition, 2003.

[MS05] Albert Meixner and Daniel J. Sorin. Dynamic verification of sequential consistency. *SIGARCH Computer Architecture News*, 33(2):482–493, 2005.

[MSB+05] Milo M. K. Martin, Daniel J. Sorin, Bradford M. Beckmann, Michael R. Marty, Min Xu, Alaa R. Alameldeen, Kevin E. Moore, Mark D. Hill, and David A. Wood. Multifacet's general execution-driven multiprocessor simulator (GEMS) toolset. *SIGARCH Computer Architecture News*, 33(4):92–99, 2005.

[MSS99] Joao P. Marques-Silva and Karem A. Sakallah. GRASP: a search algorithm for propositional satisfiability. *IEEE Transactions on Computers*, 48(5):506–521, 1999.

[NdPC+04] Jose A. Nacif, Flavio M. de Paula, Claudionor N. Coelho, Fernando C. Sica, Harry Foster, Antonio O. Fernandes, and Diogenes C. da Silva. The Chip is Ready. Am I done? On-chip Verification using Assertion Processors. In *Symposium on Integrated Circuits and System Design*, pages 55–59, September 2004.

[PBG05] Mukul R. Prasad, Armin Biere, and Aarti Gupta. A survey of recent advances in SAT-based formal verification. *STTT, Journal on Software Tools for Technology Transfer*, 2:156–173, April 2005.

[PF05] Douglas L. Perry and Harry Foster. *Applied Formal Verification: For Digital Circuit Design*. McGraw-Hill Professional, first edition, 2005.

[Piz04] Andrew Piziali. *Functional verification coverage measurement and analysis*. Springer, first edition, 2004.

[Rot00] Hemant Rotithor. Post-silicon validation methodology for microprocessors. *IEEE Design and Test of Computers*, 17(4):77–88, October 2000.

[RS95] Kavita Ravi and Fabio Somenzi. High-density reachability analysis. In *ICCAD, Proceedings of the International Conference on Computer Aided Design*, pages 154–158, February 1995.

[SC04] Tom Shanley and Bob Colwell. *The Unabridged Pentium 4: IA32 Processor Genealogy*. Addison-Wesley Professional, illustrated edition, 2004.

[SFH+03] Isic Silas, Igor Frumkin, Eilon Hazan, Ehud Mor, and Genadiy Zobin. System-level validation of the Intel® Pentium® M processor. *Intel Technology Journal*, 07:38–43, May 2003.

[SKC95] Bart Selman, Henry Kautz, and Bram Cohen. Local search strategies for satisfiability testing. In *DIMACS Series in Discrete Mathematics and Theoretical Computer Science*, pages 521–532, 1995.

[SNC+07] Smruti Sarangi, Satish Narayanasamy, Bruce Carneal, Abhishek Tiwari, Brad Calder, and Josep Torrellas. Patching processor design errors with programmable hardware. *IEEE Micro*, 27(1):12–25, 2007.

[Som09] Fabio Somenzi. *CUDD: CU Decision Diagram Package*, 2009. `http://vlsi.Colorado.edu/~fabio/CUDD`.

[SV06] Sean Safarpour and Andreas Veneris. Abstraction and refinement techniques in automated design debugging. In *MTV, Proceedings of the International Workshop on Microprocessor Test and Verification*, pages 88–93, December 2006.

[SVM07] Sean Safarpour, Andreas Veneris, and Hratch Mangassarian. Trace compaction using SAT-based reachability analysis. In *ASP-DAC, Proceedings of the Asian-South Pacific Design Automation Conference*, pages 932–937, January 2007.

[Syn09] Synopsys®Inc. *VCS: Comprehensive RTL Verification Solution*, 2009. `http://www.synopsys.com/tools/verification/functionalverification/pages/vcs.aspx`.

[VR05] Srikanth Vijayaraghavan and Meyyappan Ramanathan. *A practical guide for system Verilog assertions*. Springer, illustrated edition, 2005.

[WBA06] Ilya Wagner, Valeria Bertacco, and Todd Austin. Shielding against design flaws with field repairable control logic. In *DAC, Proceedings of the Design Automation Conference*, pages 344–347, July 2006.

[WBA07] Ilya Wagner, Valeria Bertacco, and Todd Austin. Microprocessor verification via feedback-adjusted Markov Models. *IEEE Transactions on Computer-Aided Design of Integrated Circuits and Systems*, 26(6):1126–1138, 2007.

[WGR05] Bruce Wile, John C. Goss, and Wolfgang Roesner. *Comprehensive functional verification the complete industry cycle*. Morgan Kaufmann, illustrated, annotated edition, 2005.

[Yer06] Siva Yerramilli. On the need for convergence between design validation and test. In *ITC, Proceedings of the International Test Conference*, page 14, October 2006.

[YPA06] Jun Yuan, Carl Pixley, and Adnan Aziz. *Constraint-Based Verification*. Springer, first edition, 2006.

Part II
FUNCTIONAL POST-SILICON VERIFICATION

This part of the book investigates emerging approaches in functional post-silicon validation of modern processors. In recent years, diagnosis of functional errors leveraging prototype hardware has been attracting interest from both research and development groups. This is primarily due to high execution performance achievable by post-silicon validation methods, compared to pre-silicon simulation techniques. However, the observability of a design's internal state is severely limited during the post-silicon phase, making it extremely difficult to identify the root cause of a bug. Consequently, approaches in this domain tend to rely on observation of errors through frequent inspection of the architectural processor state via silicon state acquisition techniques. We begin the first chapter in this part of the book with a detailed overview of post-silicon validation, as it is conducted in the industry today, and identify a critical bottleneck in this traditional framework. Subsequently, we discuss a novel solution that eliminates this bottleneck, unlocking the full potential of this methodology. In the following chapter we focus our discussion to multi-core designs and first examine the typical architectural framework for these devices and identify shared memory communication as the primary challenge in the verification of such designs. Three solutions for post-silicon validation of the memory subsystem in chip multiprocessors are then presented in that and the following chapter.

Chapter 3
POST-SILICON VALIDATION OF PROCESSOR CORES

Abstract. Verification remains an integral and crucial phase of the modern microprocessor design and manufacturing process. Unfortunately, with soaring design complexities and decreasing time-to-market windows, today's verification approaches are incapable of fully validating a processor design before its release to the public. Increasingly, post-silicon validation is deployed to detect complex functional bugs, in addition to exposing electrical and manufacturing defects. This is due to the significantly higher execution performance offered by post-silicon methods, compared to pre-silicon approaches. We begin this chapter with an overview of traditional post-silicon validation techniques, as reported by the industry. We pay special attention to error detection and debugging methodologies discussed in the literature and identify several crucial drawbacks in traditional post-silicon techniques. In particular, we show how the performance of architectural simulators, used to determine the correctness of post-silicon tests, have become a bottleneck in current methodologies. We then discuss in detail a novel solution to address this issue, called *Reversi*. Reversi generates random programs in such a way that their correct final state is known at generation time, thus completely eliminating the need for architectural simulation. At the end of the chapter, we demonstrate experimentally that Reversi generates tests exposing more bugs faster, and can speed up post-silicon validation by 20 times, when compared to traditional flows.

3.1 Traditional post-silicon validation in industry

Verification remains an unavoidable, yet quite challenging and time-consuming aspect of the microprocessor design and fabrication process. With shortening product timelines and increasing time-to-market pressure, processor manufacturing houses are forced to pour more and more resources into verification. The problem is exacerbated by the appearance and growing adoption of multi-core architectures. The design effort in these systems is often less than that of a single-core processor of similar size, since here identical cores are replicated multiple times within the design,

I. Wagner and V. Bertacco, *Post-Silicon and Runtime Verification for Modern Processors*,
DOI 10.1007/978-1-4419-8034-2_3, © Springer Science+Business Media, LLC 2011

occupying a large fraction of the silicon die area. The verification effort is higher however, because inter-core communication must also be verified, in addition to the cores themselves; and it often introduces non-deterministic behavior in the system as a whole. Therefore, with processor complexity increasing rapidly, and verification coverage lagging behind, bugs, such as the AMD Opteron REP MOVS error [AMD05] and the functional problems in the Intel's Core 2 Duo [Int07a, Int07b], continue to slip into production silicon.

As discussed in the previous chapter, hardware verification can be divided into three phases: pre-silicon, post-silicon and runtime. Pre-silicon verification employs two major families of solutions: simulation-based tools and formal techniques. Although formal solutions can be used to prove key design properties, such as absence of deadlock, proper ALU and FPU functionality, *etc.*, they suffer from the memory explosion problem and can ultimately be used only on small design modules. For example, in the verification of the Intel Pentium 4 processor, formal methods were used only on floating-point units, schedulers and instruction decoders [Ben01]. Simulation approaches, on the other hand, do not have such strict limitations, but neither can provide hard guarantees of correctness: only those behaviors that have occurred during the simulation can be validated. Nevertheless, simulation remains the method of choice for pre-silicon verification due to its scalability.

Post-silicon validation relies on a concept similar to simulation: the hardware prototype executes as many randomly generated input vectors as possible. However, there are a few key differences between this approach and pre-silicon verification. First, the execution on a hardware prototype is several orders of magnitude faster than any functional simulator, therefore, significantly more test vectors can be checked. However, this high speed comes at the price of limited observability: the internal state of the prototype cannot be fully and easily observed, forcing engineers to diagnose errors from the architectural state of the system or to periodically suspend test execution to record the internal state of the processor via special testing interfaces. Finally, in addition to functional testing, post-silicon validation allows engineers to test electrical properties of the design, such as maximum frequency, compliance with I/O interface specifications, *etc.*

Runtime solutions, which have started to appear recently, take verification past the release of the product and ensure correctness of operation of the device in the field. To this end, the device is typically augmented with extra hardware to monitor the execution, detect errors and violations of invariants and recover from them. In addition to avoiding functional bugs, runtime verification approaches can also provide protection from transient errors, which occur because of energetic particle strikes, and permanent faults due to transistor wearout. In the later chapters of the book we investigate such runtime solutions in much detail, so let us now concentrate on post-silicon techniques employed by the industry.

As reported in [Rot00, SFH$^+$03, BASS07, KAI07], industrial-scale post-silicon validation can divided into two parallel thrusts: *compatibility verification* and *system validation* (Figures 3.1 and 3.2, respectively). The former one analyzes the operation of the system in a typical execution environment, such as a standard desktop or server system. Thus, for testing purposes, the prototype is inserted into a reference

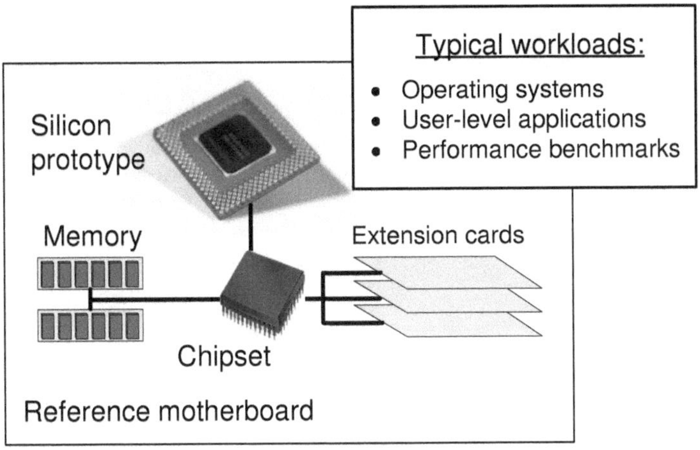

Fig. 3.1 Post-silicon validation: compatibility verification setup. The processor's silicon prototype is inserted into a reference motherboard with commercially available chipset, memory and extension cards. The system then runs directed tests such as operating systems, user-level applications and performance benchmarks. While compatibility verification has relatively low cost, its coverage of the newest and, thus, most bug-prone features is low, because the software applications do not extensively probe these components.

motherboard equipped with off-the-shelf available hardware. The design is then exercised by running a variety of commercial applications, such as operating systems, legacy software, performance benchmarks, and so on. The advantage of compatibility verification is that it is relatively cheap, since both the software tests and the hardware infrastructure (that is, the platform on which the system is running) are, for the most part, readily available and mature components. Moreover, error detection is conducted in a fairly straightforward fashion: bugs can be detected through a system hang, application error messages, *etc.* In other words, engineers leverage the known-correct behavior of the hardware and software platform in compatibility testing as the yardstick to determine processor correctness. The drawback of this type of post-silicon validation is that it provides very low coverage of the new processor features, since those are not stimulated by legacy applications. Furthermore, since tests are executed on an average-user platform, the "stressfulness" of the validation stimuli is also low, leading in turn to insufficient coverage. Finally, debugging during compatibility verification is extremely complex, being exacerbated by the lack of specialized features to suspend test execution and provide access to internal processor signals' values.

System validation, on the other hand, is typically carried forward through randomized programs, which attempt to thoroughly investigate the broad set of possible operations of the processor, including corner case behaviors, as well as high traffic and high computation density operations. In this setup, quick error detection and debugging are greatly emphasized, as is the need for coverage metrics, *i.e.,* the measure of effectiveness of a test program in stimulating certain aspects of the de-

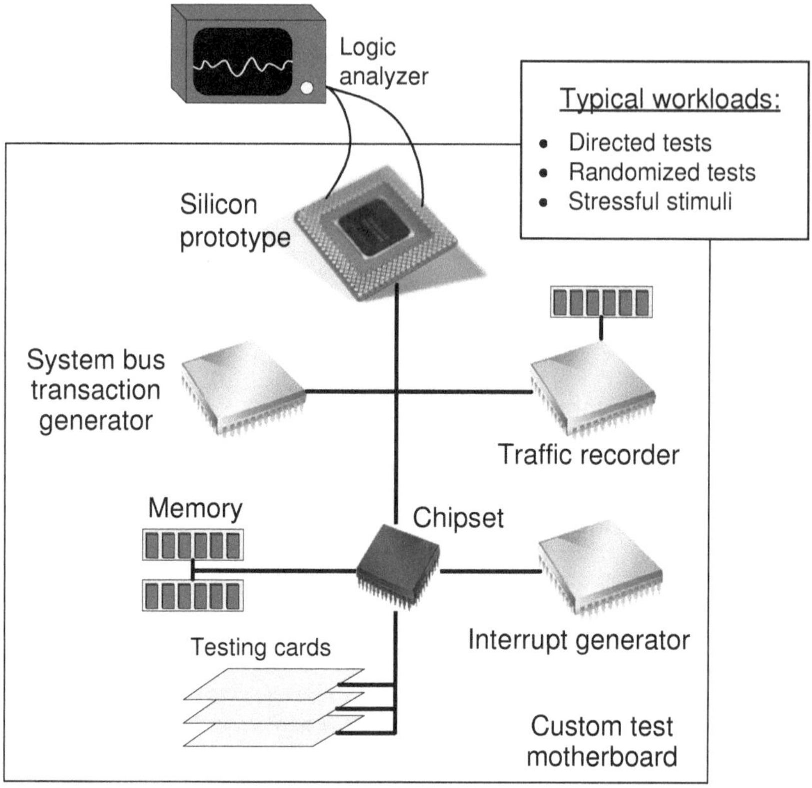

Fig. 3.2 Post-silicon validation: system validation setup. The prototype silicon processor is verified on a custom-built motherboard equipped with a number of testing and debugging modules, including tester cards, off-chip communication traffic snoopers/recorders, *etc.* In system validation, the typical workload includes stressful randomized programs targeting hard-to-reach corner cases of the design's behavior. In addition, new features and components are highly stimulated in an attempt to expose subtle bugs. Note that, given the randomized nature of these tests, system validation must rely on off-chip architectural level simulation to determine the correct test outcomes.

sign to expose potential flaws. To this end, system-level validation platforms are often custom-built by the product verification teams and are equipped with special additional hardware for testing and debugging, such as off-chip traffic snoopers. In addition, they provide connectivity to external logic analyzers, and the ability to run programmable test generators, as shown in Figure 3.2. An example of a real-world system validation setup is presented in a work discussing post-silicon validation of the Pentium M processor [SFH+03]. There, the validation platform included an interrupt generator (dubbed the "mobile agent") and a system bus transaction generator. The former was to assert certain signals, such as external interrupts, and emulate a variety of system-wide events in response to triggers sent out by the processor. The transaction generator was designed to observe and dynamically modify the data traffic between the processor and the chipset. It could respond to bus requests with predefined or randomized data and subject the processor to stressful activity. Both these components were implemented on a programmable FPGA fabric, and thus could be easily adapted to focus on the validation of a range of different processor features. In addition, for some tests, they could be used in lieu of memory and external peripherals and emulate their interaction with the processor. Finally, the testing board included a traffic recorder to log the activity observed on the external processor bus for debugging purposes.

A more recent example of an industrial-scale system validation platform, in this case for the Intel Core 2 Duo design, was reported in [BASS07]: the authors augmented the platform with a high-speed logic analyzer, configured to record system bus transactions for offline error checking and coverage analysis. For the first objective, the observed data was sent to batch servers that checked it using predefined protocol-level rules. The coverage measurement was provided with the aid of a database of all known system bus transactions and all possible pairs of back to back transactions. After each successful test, observed bus messages were tagged in the database until the desired coverage level was achieved (for this particular product it was 100% coverage of single transactions and 54% of all pairwise combinations). Although these techniques provide great support to the validation process, note that all of them operate at the *architectural level*, observing and controlling the processor only through its I/O pins, and keeping the internal state and activity of the system obscured from the engineers. Therefore, in most cases, bug diagnosis is extremely challenging using these methods, since basic information of internal processor activity is missing. Furthermore, even the coverage information that can be collected is extremely partial and based on the little data that can be collected from external observation, with the result that high-quality post-silicon validation is very hard to achieve with this type of frameworks.

The drawback of limited internal observability of a processor prototype can be addressed by probing or other silicon state acquisition techniques, as outlined in Section 2.2. Recall that optical probing methods can be deployed for sampling of individual wires and transistors on a silicon die [EWR00], but tend to be inefficient since only a handful of locations can be monitored at once. For more efficient diagnostic, industrial designs are often augmented with special on-die features, such as scan-chains [KDF+04], JTAG ports [MDM+06], cycle breakpoints [BEC06] and on-chip logic analyzers [Lit02], which allow high-speed access to much of the internal state of the device. As a tradeoff, however, these techniques incur some area penalty and are limited in the amount of information that can be observed "at speed". Thus, after the internal buffers of these components become full, execution must be suspended and the recorded data must be transferred to external storage. Recently, a more homogeneous framework for post-silicon state acquisition was proposed in [ABD+06], and so far it has been employed in several industrial validation projects.

In the context of microprocessor verification, these general solutions are often complemented by domain-specific techniques to further improve observability and accelerate the debugging process. For instance, as discussed in [SFH+03], during the validation of the Intel's Pentium M processor the verification team leveraged a periodic state management interrupt (PSMI) mechanism to access the microarchitectural state of the processor. When the PSMI interrupt pin was asserted, execution was suspended and a software routine would upload the internal state of the processor to memory. By pulsing this signal periodically, the authors obtained several checkpoints close to the error manifestation time and used them to replay tests and diagnose the root cause of the bugs. Internal state observability can also be used to improve the quality and the data collection process of post-silicon coverage, as indicated in [BASS07]. In that work, the authors set internal processor counters to monitor the occurrence of certain events during execution. By means of these counters, rare microarchitectural events could be identified, and the team could parametrize test generation to target the corresponding insufficiently validated scenarios. Another interesting coverage data collection technique outlined in [BASS07] relies on the microcode patching mechanism often available in modern processors. We will overview microcode patching in a later chapter, along with other runtime solutions; in [BASS07] it was not employed to fix errors, but instead to detect occurrences of internal events. Specifically, the authors developed custom microcode patches to be invoked when certain instructions were in execution; these patches would sample the processor's state and transmit it to the external platform through JTAG ports. This solution enabled a significant increase in the amount of microarchitectural information collected and greatly improved post-silicon validation coverage. Similar experiences were reported by the multi-core UltraSPARC T2 post-silicon validation team [KAI07]. In this project, debugging features such as scan-chains, on-chip triggers, and JTAG ports were exploited to observe the prototype's internal state, while the test generation was conducted in software running on the processor itself. Particular attention in this case was dedicated to the validation of the inter-core communication block and a novel on-die encryption module included in the T2 design.

The design of the framework used for obtaining the silicon state, however, is just one aspect of post-silicon debug, since even when access to the system's internal signals is available, it is often extremely hard to diagnose the root cause of an issue. This is due to the extreme complexity of modern microprocessors, where the effects of an error are often manifested millions of cycles after its occurrence. Therefore, engineers often spend much of their time replaying the test and sampling the internal state at different points in time and with different time granularities, ideally "zooming in" on the bug. Supporting the engineering team in this arduous task are trace minimization tools, such as [CBM07b]. This solution accepts as inputs a simulation trace exposing a functional error and structural information about the circuit, producing a much shorter test sequence that still reveals that same bug. When short bug traces are available, it is possible to perform debugging using RTL simulation, leveraging methodologies similar to pre-silicon verification, where observability and controllability are not an issue. After the source of the error is identified, the validation team investigates how it can be corrected and suggests RTL modifications to be implemented in later hardware versions. Overall, as you may imagine, the process of post-silicon debugging and repair requires significant insights into the functionality and implementation of the circuit-level structures, necessitating close human guidance. A recent research solution in [CWBM07] proposed to automate this process with the use of formal verification techniques: given a buggy RTL design, a high-level (architectural) golden model and a trace leading to the error, the tool presented in that work automatically augments the design with additional internal signals and then formulates the diagnosis problem as a pseudo-Boolean set of constraints. Solving this problem allows the technique to pinpoint the location(s) in the RTL code that are responsible for the incorrect behavior and to suggest modifications that will repair the error. Notice that this approach does not require internal design state information, and only needs a set of correct output responses to test inputs. However, since the correct system behavior in this case is defined only by these responses, the tool will only suggest modifications that make the outcomes of the RTL and architectural models match for the input sequences provided, and cannot guarantee system's correctness for other test sequences. While promising, this technique may not be scalable to circuits as complex as microprocessors and, to the best of our knowledge, has not been deployed in industry contexts, where post-silicon debug still remains more of an art, than a scientific methodology.

In summary, a few concepts and challenges are common to all post-silicon validation methodologies discussed so far: i) observability and controllability are scarce, ii) raw system simulation performance is very high, iii) diagnosis and correction are extremely challenging, and to date very few systematic methodologies are available to attack this problem. Finally, iv) all system-level validation activities require the use of *randomized tests*, which must be capable of cleverly stimulating different design activity to boost the validation coverage. While these program sequences allow the hardware prototype to be stressed, that is, be subjected to intense activity in one design aspect, nevertheless, they have an inherent handicap in their lack of test outcome knowledge. In other words, when a random test is generated, designers do not know what the correct result of its execution should be, and are forced to run this program on a distinct golden model of the same design. The final state of the golden model can then be compared to the state of the hardware prototype, which is obtained via silicon access mechanisms, as discussed above. To provide correct responses, two types of golden models can be used: if the test only exercises legacy instructions, an older generation processor can provide the correct outcome; on the other hand, to validate new features of the design, engineers must resort to logic simulation or, possibly, emulation. Clearly, validation of the new features is of critical importance, since often new components are the source of the majority of bugs. As a consequence, simulation, which even at the architectural level performs orders of magnitude slower than actual silicon, becomes the bottleneck in this process. Under these circumstances, design houses are forced to spend enormous computational resources on server farms, in charge of simulating many tests in parallel, trying to keep up with the high performance of the silicon prototype [Rot00]. If, however, the final outcomes of these tests were to be known *a priori*, as it is in directed tests, it would be possible to avoid this simulation bottleneck, and greatly speed up random-testing post-silicon validation, while keeping the benefits of its high coverage. A solution designed exactly for this purpose was proposed recently by us in [WB08]. In the remainder of this chapter we analyze this technique in detail, presenting an example of generated test programs and evaluating the performance of the approach.

3.2 The Reversi Test Generation System

Typically post-silicon functional validation in industry has been conducted with two types of tests: parameterized directed tests and constrained-random (or pseudo-random) tests. Although the former ones can provide high coverage, they require significant human effort to be developed. Pseudo-random tests, constrained to produce only valid instruction sequences, can be generated automatically, however, often suffer from lower coverage. More importantly, the final state of the processor after executing a random test sequence is unknown. Therefore, engineers must resort to simulating the design's golden model to compute the final processor state and check it against state dumps of the actual hardware prototype (Figure 3.3).

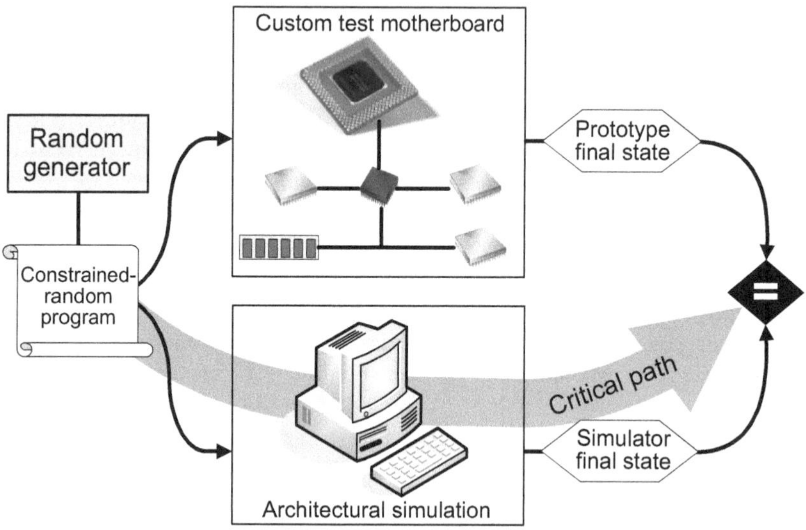

Fig. 3.3 A typical post-silicon system validation flow. In a typical post-silicon methodology, random tests are produced by a test generator and fed to both a golden model simulator and a silicon prototype. Bugs are flagged by differences between the prototype's and simulator's final states. Both test generation and simulation are performed by software executing on a host machine at relatively low performance when compared to the test execution performance of the silicon prototype. Thus, the performance bottleneck of this setup is the architectural simulation server.

Unfortunately, the simulation of the golden model is several orders of magnitude slower than the hardware execution, therefore, the computation of the final state becomes a bottleneck for the entire effort. The methodology proposed here addresses this issue by developing a post-silicon solution which fully exploits the performance of the hardware under test. It is based on a test generator, called *Reversi*, that produces tests whose outcome is known by construction. This allows Reversi-based post-silicon flow to bypass the simulation step and speed up the overall validation, as illustrated in Figure 3.4.

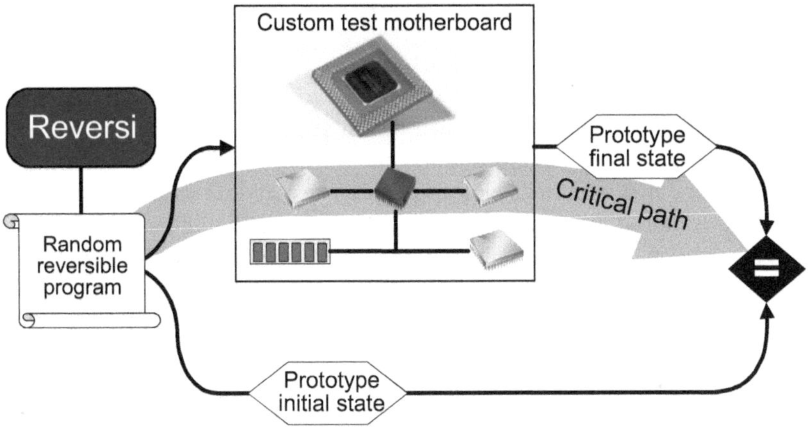

Fig. 3.4 A Reversi post-silicon system validation flow. A Reversi-based flow does not require an architectural simulator: random reversible programs can be generated on a tester board or on the hardware prototype itself. These programs are characterized by producing matching initial and final processor state when executed correctly. Consequently, bugs are flagged when differences are observed between the initial state of the prototype and its final state after executing the test.

The key observation that allowed the development of Reversi is that many instructions in a processor's ISA have counterparts, *i.e.,* operations whose functionality has a corresponding inverse instruction, such as restoring a value in a particular register, clearing a set of flags, *etc.* Moreover, if no single instruction exists to reverse the action of another, one can devise a small program sequence to be used to the same effect. This was the case, for example, for the integer multiply instruction in one of the ISAs that we will discuss in the experimental evaluation presented in Section 3.5. No instruction for integer division was available in that ISA, thus software emulation of division was used to revert the effects of multiplication. Note that, if the emulation routine exposed any error in the hardware prototype, the result of the multiplication would not be reversed correctly. The presence of inverse functions enables generation of programs that include every instruction in an ISA, and for which the final register values match exactly the initial ones. In other terms, if $\bar{x}$ is a vector representing the processor state, and each F_i / F_i^{-1} pair represents a distinct function (either an ISA instruction or an instruction block) and the corresponding inverse, then a program generated by Reversi can be thought of as the following sequence: $F_1^{-1}(F_2^{-1}(...(F_n^{-1}(F_n(...(F_2(F_1(\bar{x})..)$. Errors in the underlying hardware can then be discovered if:

$$\bar{x} \neq F_1^{-1}(F_2^{-1}(...(F_n^{-1}(F_n(...(F_2(F_1(\bar{x})..) \tag{3.1}$$

3.3 Reversible and Non-reversible Instructions

In order to create reversible programs, one need to first analyze the design's ISA and identify inverse instructions (or instruction sequences) for each of the operations. By applying these operations in the manner discussed above, the state of the processor can be first modified and then properly restored (in the absence of bugs). This can be achieved by creating a *block database* containing pairs of *functional blocks*: for each operation block, there is a corresponding inverse block. Each block contains either a single instruction or a small program sequence. An operation block modifies the value of a register, called the *focus register*, while its inverse restores its initial value. The ID of the focus register for each block is a parameter set by Reversi dynamically during test generation. Therefore, the same block may appear in the test program multiple times, each time modifying a different register, which allows a varied set of programs to be created. Note that blocks operate only on a single focus register at a time to maintain the reversibility of the program and track the correctness of its execution. Thus, for instructions with multiple operands, only one register is the focus register, while other operands are randomly generated by Reversi according to the instruction format. The flexible and robust structure of the block database allows the Reversi algorithm to be agnostic to the functionality of individual blocks and the underlying ISA, making the framework readily adaptable to different processor architectures. Moreover, since each block in Reversi may contain multiple instructions, it is possible to populate the database with complex functions, including loops, procedure calls, *etc.*, and create elaborate tests representative of real software applications. In the following section we discuss individual classes of instructions and details of operation and inverse block verifying them.

3.3.1 Arithmetic and logic instructions

The design of blocks containing arithmetic and logic instructions is summarized in Table 3.1 and is fairly straightforward, since the majority of these operations have a simple inverse directly available in the ISA. For example, *add* (addition) can be reversed by *sub* (subtraction), while *inc* (increment) can be reversed by *dec* (decrement), *ror* (rotate right) by *rol* (rotate left) and so on. If an instruction does not have a counterpart in the ISA, a small routine can be used to emulate its inverse. Some Boolean logic instructions, such as *and* and *or*, do not have direct inverses, however, these operations can be used to construct an *xor* logic function, which can then be reversed by another *xor* instruction. Such structure is also beneficial for verification of the *xor* instruction itself, since the operation and its inverse block in this case exercise different hardware modules. Situations where the same processor components are exercised to execute a function and its inverse should be avoided this is the last candidate. next esc will revert to uncompleted text. o prevent bugs being masked by faulty hardware.

Table 3.1 Reversi blocks for arithmetic and logic instructions.

Instruction	Operation Block	Inverse Block
add	*add*	*sub*
sub	*sub*	*add*
inc	*inc*	*dec*
dec	*dec*	*inc*
xor	*xor*	*xor* function emulated by a combination of *and*, *or*, and *not* instructions: a *xor* b = (a *and not*(b)) *or* (*not* (a) *and* b)
not	*not*	*not* function emulated with *nand*: *not* a = a *nand* a
neg	*neg*	negation function emulated with multiplication by -1: -a = a*(-1)
and	*xor* function emulated by a combination of *and*, *or*, and *not* instructions: a *xor* b = (a *and not*(b)) *or* (*not* (a) *and* b)	*xor*
or	*xor* function emulated by a combination of *and*, *or*, and *not* instructions: a *xor* b = (a *and not*(b)) *or* (*not* (a) *and* b)	*xor*
mult	*mult*	integer division emulated with Booth's algorithm subroutine
rol	*rol*	*ror*
ror	*ror*	*rol*
sll	two step process: 1. store lost bits 2. *sll*	two step process: 1. *srl* 2. restore lost bits
srl	two step process: 1. store lost bits 2. *srl*	two step process: 1. *sll* 2. restore lost bits
sra	three step process: 1. store lost bits 2. create mask 3. *sra*	three step process: 1. *rol* 2. apply mask 3. restore lost bits

Some ISA operations, for example *sll* (logic shift left) and *srl* (logic shift right), cause some of the data bits to be lost. In order to be able to restore fully the initial value of the focus register, we must mask out these bits and store them in a scratch-pad memory before applying the operation. When the program reaches the inverse block, it first applies the reverse operation (*i.e.,* shift in the opposite direction in this case) and then loads and restores the bits from memory. Finally, the outcome of an

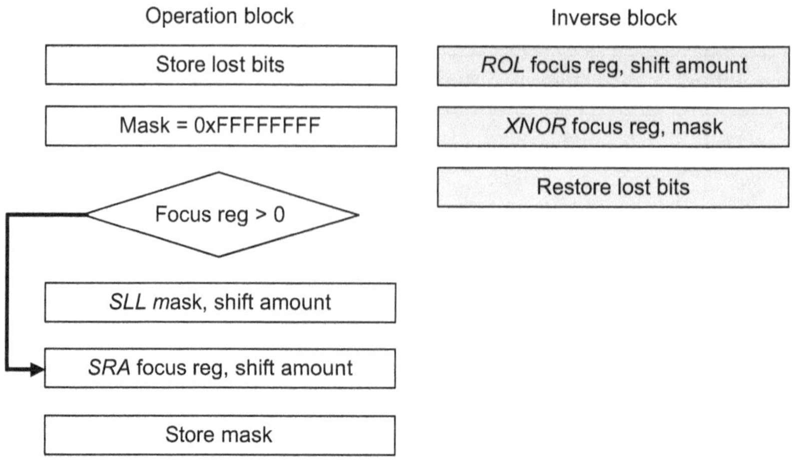

Fig. 3.5 Operation and inverse blocks for the shift-arithmetic right *sra* instruction. The operation block first stores aside the bits that would be lost when executing the *sra* instruction. Then a filter mask is set up based on the arithmetic sign of the focus register; the *sra* is executed, and finally the filter mask is also stored aside. The inverse block uses a rotate left instruction to relocate the bits of the focus register in their original position, then uses the mask to filter sign bits that were set during the *sra* instruction, and finally restores bits that were eliminated by the shift operation.

instruction may depend on the sign or value of the focus register, which is not known at generation time. For example, shift-arithmetic right (*sra*) will preserve the sign of the value by replicating its most significant bit. To detect errors in this operation we developed the scheme illustrated in Figure 3.5: first, the bits that will be lost due to the shift operation are stored in memory, then the sign of the focus register is checked and a bit-mask is created accordingly. For positive values of the register, the mask is *0xFF..F*, while for negative values the mask has the bits corresponding to the shift amount (*SA*) set to 0 (note that the shift amount is encoded in the instruction itself, thus known at generation time). Finally, the *sra* function is applied. To check the correct operation of the function in the inverse block we first apply a (rotate left) *rol* instruction, and then overlay the mask with an *xnor* operation. The use of a rotate operation in the inverse block allows us to detect bugs in the *sra* instruction that affect the most significant bits of the focus register. The subsequent *xnor* operation will propagate potential bugs in any bit position of the register. To better illustrate this process, we present an example of a short program generated using this scheme for two systems in Figure 3.6. The left part of the image shows an assembly-level program generated by Reversi from the template in Figure 3.5. The system in Figure 3.6.b implements the operation correctly, and thus executes the program correctly, restoring the original value of the focus register, while the other system does not replicate the sign bit properly due to a bug in its hardware implementation, an issue detected by Reversi through the mismatch of initial and final focus register's values. Errors in the arithmetic shift operation for positive values of the focus register will also be detected by programs generated following this template.

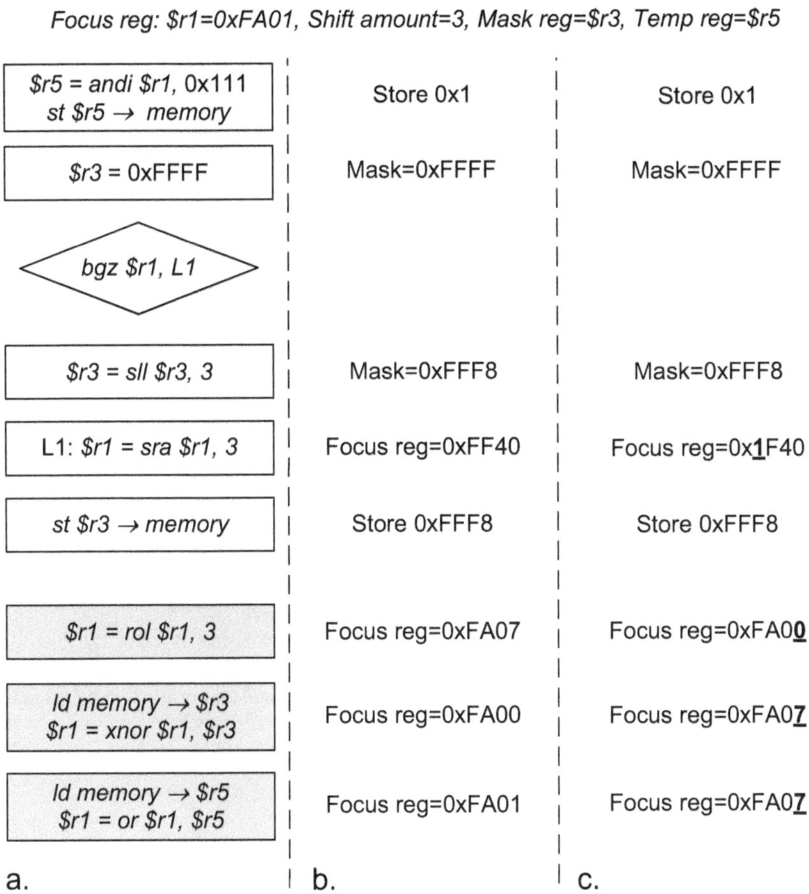

Fig. 3.6 An example of a program testing the shift-arithmetic right *sra* instruction. The left part of the figure shows an example program generated based on the template of Figure 3.5. Execution on two systems is also shown: the system on the left (**b.**) is implemented correctly, and thus restores the original value of the focus register upon completion of program execution. The system on the right (**c.**), on the other hand, includes an implementation bug, which manifests when executing the *sra* instruction. The erroneous value computed in the focus register is propagated through the program and can be detected at completion by simply comparing the initial and final values of the focus register.

3.3.2 Load/store instructions

In Reversi, the correctness of load and store instructions is checked by copying a data structure to and from memory: a region of memory is initialized with random values and load/store pairs are used to copy it to a new location. The copy operation is not required to preserve the order of the data bytes, rather, the data structure is treated as a pool of values, which can appear out of order at destination, as illustrated in the schematic of Figure 3.7. This allows tests generated by Reversi to closely resemble real applications, where load instructions bring data from memory to the processor, and stores copy results of the computation back to memory. Moreover, because of their random nature, Reversi programs contain a variety of cache and memory access patterns, that can expose corner case bugs in the memory subsystem. To check the correctness of the final state of the memory, a signature of the memory values is obtained by hashing them together using *xor* instructions. This signature is computed at the beginning and at the end of the test and the two values are compared to detect possible mismatches. This approach allows Reversi to expose load/store-related issues such as illegal memory accesses and/or data corruption.

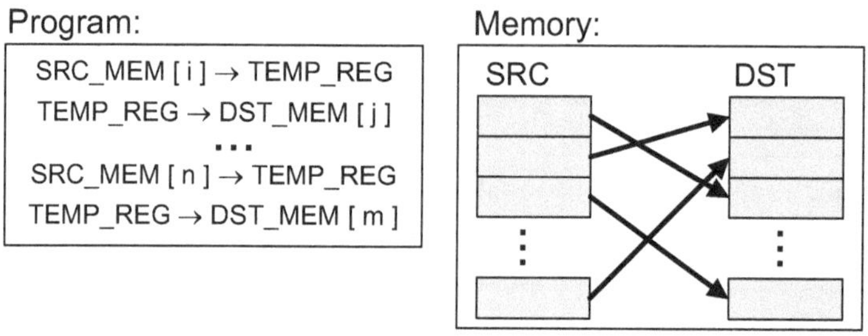

Fig. 3.7 Testing of load and store instructions in Reversi. Load/store instruction pairs are used to copy bytes between two memory regions. The order of data values may be reshuffled, but values must not be eliminated nor duplicated, since their *xor*-based signatures must match.

3.3.3 Branch instructions

A basic block database of Reversi would also contain templates to test a range of branch instructions with different properties: forward/backward branches, taken/not-taken, *etc.*For example, an operation block for a forward taken branch (Figure 3.8) contains a store operation that saves the value of the focus register to scratchpad memory, followed by a load that overwrites the focus register with a predetermined constant. Finally, the branch instruction itself follows, branching conditionally on the focus register holding the constant value just loaded. Although the constant is generated randomly, its value is dependent on the type and specific instance of the tested branch, to support bug diagnosis in case of test failure. For example, a template for a *beq* (branch if equal) instruction overwrites the focus register with a random constant and then loads a temporary register with the same value used in the branch instruction under test. With reference to Figure 3.8, the destination of the branch is located in the inverse block, which also contains a load operation restoring the focus register and a return (to the operation block) jump. Therefore, if all control flow instructions are executed correctly, the value of the focus register after the execution of these blocks is preserved. If, however, the branch is not taken by a faulty hardware, the focus register would not be restored. Note that the code in the inverse block is only accessible via the corresponding branch under test and it is bypassed otherwise. On the other hand, if the unconditional branch fencing off the inverse block of Figure 3.8 is not taken due to a bug, the *halt* instruction is executed, and the test ends without fully restoring the original processor state.

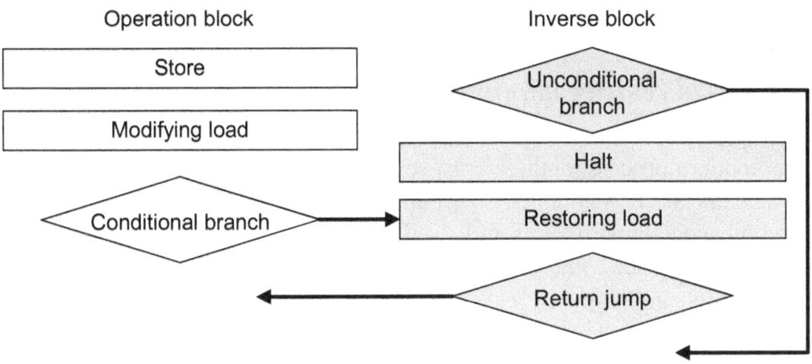

Fig. 3.8 Reversi template for forward-taken branch instructions. The operation block includes a load instruction modifying the value stored in the focus register. The load is preceded by a store instruction to preserve the current focus register's value. A conditional forward branch testing for the value just loaded in the register is then the central part of this operation block. If the instruction executes correctly, the branch will be taken, execution will transfer in the inverse block, where the saved value of the focus register is restored, and then execution is returned to complete execution of the operation block. The inverse block should be simply skipped in a fully functional processor, as indicated by its first instruction being an unconditional jump.

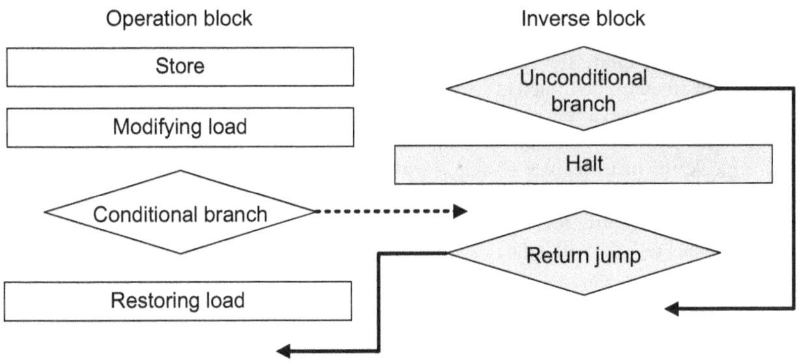

Fig. 3.9 Reversi template for forward not-taken branch instructions. The operation block operates similarly as the one in Figure 3.8; however, now the conditional branch should not be taken, since the focus register is tested against a different value from what was just loaded, and the restoring load is part of the operation block now. In case of execution of this flow in a faulty hardware, it is possible that the branch is still taken erroneously, in which case the restoring load would be bypassed by the chain of jump instructions, and the final value of the focus register would be incorrect, revealing the bug.

The structure of the blocks for a forward not-taken branch is similar to the one described above and it is shown in Figure 3.9, differing only in the location of the restoring load. Templates for other types of control flow instructions can be inferred from the examples shown in Figures 3.8 and 3.9. We use a similar technique to detect other faulty control flow operations, initializing all unused locations in the program to *halt* instructions.

3.3.4 Control register manipulation

In many modern processors there exist several special control registers. In general terms, we can classify them into two groups: *mode control* registers, that can only be accessed by special instructions and specify the machine's mode of operation; and *execution flag* registers, that cannot be changed directly by the user/programmer, but are affected indirectly by executed instructions. For instance, a register enabling/disabling the first level cache is a mode control register, while a register storing the ALU *overflow* bit, or comparison result bits from the comparator (*equal, greater-than, etc.*) is an execution flag register.

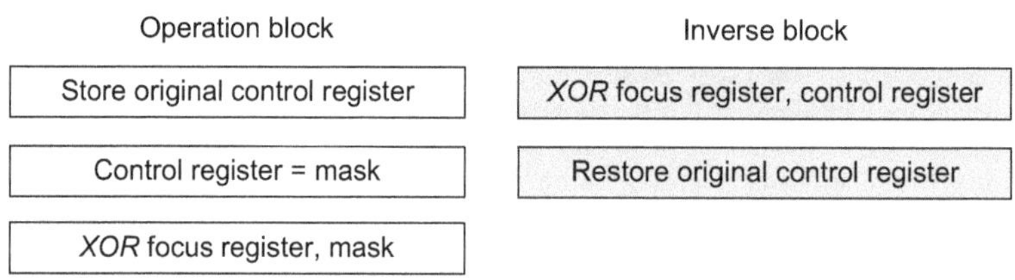

Fig. 3.10 Reversi template for mode control registers. Mode control registers are tested by exercising the instructions dedicated to set their internal bits. In the operation block a mode control register is overwritten with a new mask, after storing its original value aside in a temporary register. That same mask is also *xor*-ed in the focus register, so that this register now reflects the update. In the inverse block, the focus register's value is using the mask stored in the control register (which should match the original mask), and then the original control register value is also restored. If the instruction setting the control register, or any of the intermediate instructions, erroneously affects the mask, the focus register will not be restored correctly, and the test execution will fail.

Reversi exploits the fact that the value of the mode control registers can be modified only through specific instructions, and must remain unchanged throughout other parts of program execution. To test the proper operation of a control register, the following sequence of steps is taken: in the operation block, the original value of the control register is first stored to memory, then the new bit-mask setting the new configuration is loaded into the control register and also *xor*-ed into the focus register. In the inverse block, Reversi first accesses the control register and *xor*s it with the focus register, ideally restoring its value. Then the original value of the control register is loaded from memory and set. The flow just described is illustrated in Figure 3.10. Note that if the control register were erroneously modified between the execution of the operation block and its inverse block, the focus register's value would reflect this error.

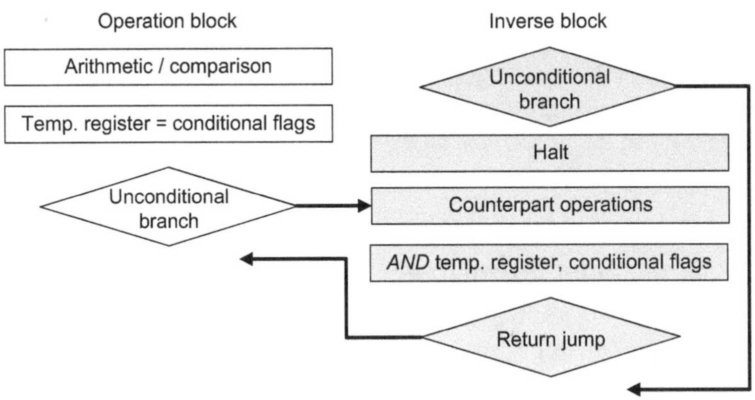

Fig. 3.11 Reversi template for execution flag registers. Execution flag registers are set as a byproduct of the execution of other instructions: arithmetic, logic, or relational. A Reversi template for these flags would first execute an instruction with the purpose of setting one of these flags, and then store the flags on a temporary register. In the inverse block a code snippet would execute to determine if the flag under test should be set or not, and then the situation is checked by analyzing the execution flag register value that was stored aside. If the flags match, execution continue from the end of the operation block. Note how in a correct execution situation, the inverse block is only accessed from the operation block.

Reversi can also check the correctness of execution flag registers, because instructions affecting them (arithmetic, logic and relational) have counterparts in terms of which flags they set. For instance, if *comp $r1, $r2* sets the *greater-than* bit, then *comp $r2, $r1* must set the *less-than* bit. Similarly, knowing that $a+b > c \equiv b > c-a$, one can check that an *add* operation sets the *overflow* bit in the execution flag register correctly. For instance, assume that

add $r1, $r2, $r3 # overflow, *i.e.,* $r3 > *MAX*

generates and overflow, because *$r3* is larger than the largest integer that can be represented in this system (*MAX*). Then one can check that the execution of this addition properly sets the *overflow* bit in an execution flag register with the following code snippet:

sub MAX, $r1, $r3
comp $r2, $r3 # must set *greater-than* bit

If the *comp* instruction sets the *greater-than* flag, then it follows that *$r2* > *(MAX - $r1)*, and thus the earlier addition must have overflowed. This example also shows that individual bits of the execution flag register can be used to check other bits in the register. The template for a Reversi block structure to validate the execution flag register is presented in Figure 3.11: the operation block executes a comparison or an arithmetic operation affecting the flags, stores the flags values in a temporary register and then jumps to the inverse block. The inverse block executes the counterpart operations, that is, it executes the necessary instructions to derive the flags values, similarly to the example provided above, and then checks that they correspond to those computed in the operation block. With reference to the example discussed, the template would first execute the *add* instruction in the operation block and then determine if it had set an *overflow* bit by executing the subtraction and comparison in the inverse block and testing the *greater-than bit*.

3.3.5 Floating point instructions

Floating point operations present a unique challenge to Reversi due to the their inherent imprecision: in the majority of cases the result of the computation is rounded, making it impossible to restore the operands exactly. To address this issue one must recognize this intrinsic approximation and take into account the relative error that is introduced by these operations. This can be done by constructing a table indexed by the operands' exponents and checking that the relative error obtained after each operation and its inverse is within the expected boundaries. Although this solution would not lead to a strictly reversible program, it is still useful for detecting bugs in the execution of floating point instructions.

3.3.6 Limitations

The Reversi framework enables the creation of high-coverage tests with a verifiable final state. However, this approach has a few limitations. The most important one stems from the fact that Reversi relies on the existence of inverse functions that can fully and precisely restore the internal state of the processor. For example, if the integer division operation were implemented in such a way that the remainder was discarded, the value of the dividend could not be restored precisely. Similarly, floating point instructions do not always provide high precision, however, they can still be partially verified by the approach just discussed. Unfortunately, input and output operations are inherently irreversible and cannot be easily validated through a Reversi solution. For example, external interrupts cannot be expected to arrive at a certain time and cannot be "undone" by the core.

In addition, Reversi is also not suitable for instructions whose output depends on the input operands' values. For instance, a divide-by-0 operation may trigger an exception and/or set an error bit. Due to the fact that input operand values in Reversi are assigned at random, it is very unlikely that a zero divisor would occur, but if it were, the block database presented in this section could not reverse the instruction properly. It is possible to address these type of issues by augmenting the database with specialized blocks to exercise corner case situations such as this.

3.3.7 Reversi Generator

As described in Section 3.2, reversible programs generated by Reversi consist of sequences of operation and inverse blocks instantiated from the block database. However, a single sequence of blocks only alters a single focus register, therefore, to create complex and more interesting programs, Reversi generates multiple block sequences (called *stacks*), each altering a different focus register. The stacks are then interleaved into a single reversible test program, as Figure 3.12 illustrates.

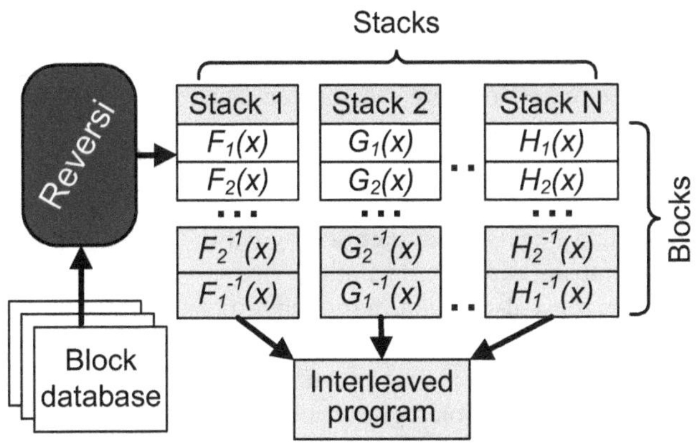

Fig. 3.12 Reversi operation. Given a database of functional blocks, Reversi produces a set of stacks, consisting of blocks and inverse operations assembled in reverse order. Each stack operates on a single focus register, modifying it in such a way that the register's final value matches the initial one. The stacks are then interleaved into a program with predictable outcome.

Stack generation. During the test generation, Reversi randomly selects functional blocks from the database and creates a user-specified number of *stacks*, each centered on a single, randomly selected focus register and consisting of several blocks and their inverses. Each stack also uses a few additional temporary registers to support the computation required in each block. Blocks are then arranged so that all inverse blocks follow all the operation blocks in inverse order (as required by Eq. 3.1). On a properly working processor the focus register should be transformed through each operation block, and then have each transformation reversed while executing the inverse block, so that at the end of the stack execution the focus register is restored to its original value. Note that the sets of temporary registers used by operations in each stack are completely disjoint to simplify stack interleaving.

Stack interleaving. After the user-specified number of stacks is generated, Reversi interleaves them by selecting instructions from each stack and chaining them together to form a single test program. Note that some instructions may be grouped together into "atomic operations", meaning that the interleaving phase cannot insert instructions between them. The atomicity indicator is provided in the block definition in the database.

To balance the selection algorithm, each stack should have a different probability of selection, based on its length, so to avoid having a long tail from a single stack at the end of the program. The probability of selecting the next instruction should be directly proportional to the remaining length of that stack, in other words the probability of selecting the next operation from a given stack j is:

$$P_j = \frac{|stack_j|}{\sum_i |stack_i|} \tag{3.2}$$

where $|stack_j|$ is the number of atomic operations in $stack_j$. After each removal of an atomic operation, all P_j's are recomputed using the updated stack lengths. Note that the requirements of using disjoint sets of registers in each stack places an upper bound on the total number of stacks that can be simultaneously interleaved in Reversi. This limitation was deemed preferable to the alternative of developing more complex dynamic register set partitioning (as in some compiler techniques) because it leads to much faster test generation.

The test program must also include a routine responsible for calculating the signatures of the values stored in the memory regions used for load/store instruction testing. This routine is executed at the beginning of the program on the source memory region, and on the destination region at the end of the test. When the program terminates, the final values of the focus registers and the signature of the destination region are compared to the initial state of the system computed and stored by Reversi in the test preamble, to determine if the execution was error-free.

It is also important to note that Reversi programs can provide more aid in debugging than traditional randomly generated programs. If the test results indicate that there is a bug in the processor, a validation engineer can quickly check if the exposing instruction sequence is located in an individual stack, by re-running the program without interleaving. Insights into the nature of the bug can also be found by "peeling" operation and inverse blocks from the program. Therefore, a reversible program exposing a bug can be dramatically shortened to alleviate debugging. In contrast, in a traditional flow, a costly re-simulation is required to obtain the new golden state after each change of the test program.

3.4 Example

This section presents an example of a program generated by Reversi for a simple instruction set outlined in Table 3.2. Two stacks for this ISA, using focus registers *$r7* and *$r11*, are shown in Figure 3.13. For both stacks the function blocks are labeled in the left column and boxes mark atomic operations. The stack on the left in Figure 3.13 contains simple arithmetic/logic operations, while the stack on the right includes logic instructions, load/store pairs and a block for a forward taken conditional branch. Sets of register IDs for both stacks are allocated dynamically by Reversi and are disjoint. Initial focus register values (*reg_val1* and *reg_val2*), constants (*const1-const3*) and locations accessed by load and store instructions in the program are also selected at random.

Table 3.2 Instruction set architecture for the example of Section 3.4

Instruction	Semantics
halt	Stop the execution
add $r1, $r2, $r3	*$r3* = *$r1*+*$r2*
sub $r1, $r2, $r3	*$r3* = *$r1*-*$r2*
neg $r1, $r2	*$r2* = -*$r1*
ld $r1, var	*$r1* = MEM[*var*]
st $r1, var	MEM[*var*] = *$r1*
beq $r1, $r2, label	PC = (*$r1*==*$r2*) ? *label* : PC+1
Register *$r0* is hardwired to the value 0	

An interleaving of the stacks into a single program is shown in Figure 3.14. Conditions that must hold after this program executes are: *$r7*= *reg_val1*, *$r11* = *reg_val2* and the signatures for source and destination memory regions used during load/store operation blocks must match. Note that the branch code in block G_2 was generated by Reversi to be taken. Thus, during correct operation, the execution should modify the value of *$r11* and jump to the label *L1*. Then the processor restores the value of the focus register and takes the unconditional branch returning to *L2*. When operating properly, the processor should not execute line *L1* again and skip directly to *L3*. If instead the branch in G_2 is not taken because of a functional error in the system, then the exit condition described above does not hold, and an erroneous value of the focus register will be propagated to the end of the execution, exposing a bug. Thus, by evaluating the final values of the focus registers *$r7* and *$r11* and the signature of the destination memory regions, Reversi can quickly determine if the program has exposed any functional bugs in the processor design.

$INIT_1$	start1: ld $r7, reg_val1
$F_1(x)$	ld $r3, const1
	add $r7, $r3, $r10
$F_2(x)$	ld $r7, const2
	xor $r10, $r7, $r15
$F_2^{-1}(x)$	ld $r5, const2
	xor $r15, $r5, $r10
$F_1^{-1}(x)$	ld $r2, const1
	sub $r10, $r2, $r7

$INIT_2$	start2: ld $r11, reg_val2
$G_1(x)$	ld $r1, src_mem1
	st $r1, dst_mem2
$G_2(x)$	st $r11, tmp_mem1
	ld $r11, const3
	ld $r30, const3
	beq $r11, $r30, L1 L2:
$G_3(x)$	ld $r1, const4
	add $r11, $r1, $r6
$G_3^{-1}(x)$	ld $r9, const4
	sub $r6, $r9, $r11
$G_2^{-1}(x)$	beq $r0, $r0, L3 halt L1: ld $r11, tmp_mem1 beq $r0, $r0, L2 L3:
$G_1^{-1}(x)$	ld $r1, src_mem2
	st $r1 dst_mem1

Fig. 3.13 Two program stacks generated for the example of Section 3.4. The stacks in the figure are generated by Reversi for the example ISA described in Table 3.2. Register *$r7* is the focus register for the stack on the left, which contains simple arithmetic and logic operation blocks and inverses. The stack on the right has focus register *$r11* and includes more complex blocks: logic, load/store and a block for a forward taken branch. In both stacks, operation blocks have been labeled in the left column, so that the correspondence with their inverse could be easily noted. Finally, each element of the stack is an atomic operation, comprising one or more ISA instructions.

3.5 Experimental Framework

To evaluate the performance of Reversi, we created two reversible instruction block databases: one implementing a subset of the DEC Alpha instruction set and another implementing a subset of the x86 ISA. The database for the Alpha instruction set contained 17 distinct blocks for arithmetic and logic functions, testing a range of instruction formats (register/register and register/immediate), and 5 blocks for each type of relational instructions. In addition to that, the database included 3 blocks for load and store instructions, an unconditional jump block and 16 branch blocks containing 4 distinct branching instructions, each in four possible modes (forward/backward and taken/not-taken).

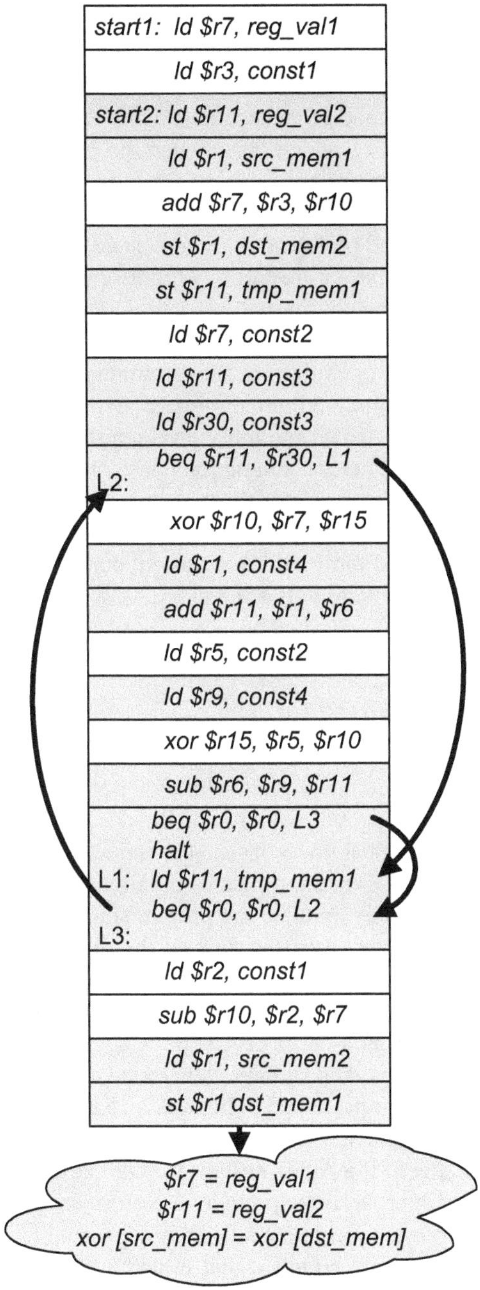

Fig. 3.14 The final test program for the example of Section 3.4. The two stacks of Figure 3.13 are interleaved using the algorithm provided in Section 3.3.7 to obtain a final Reversi-generated test program. Note that the final values of the focus registers are compared against the initial values stored in memory, and the signatures of the load/store operation blocks memory region are checked.

Similarly, the x86 block database contained 32 logic-arithmetic blocks testing multiple instruction formats (register/register, register/immediate, register/memory, memory/register), 3 load/store blocks, 1 compare block and 40 branch blocks, for a total of 76 operation blocks and 76 inverse blocks. Reversi itself was implemented as an optimized program in C, which instantiated the block database, created and then interleaved a specified number of stacks. Additionally, Reversi augmented the test program with specialized routines initializing the state of the processor with random-generated values and performing the final state check when the test terminates. The blocks of the database were partially pre-assembled in binary, and Reversi was responsible for setting the appropriate bit-fields in each instruction with register IDs, randomly generated constants, *etc.* These bit assignments were all performed on demand through bit-wise operations during program generation.

To compare Reversi with a traditional system post-silicon validation setup as shown in Figure 3.3, we created an assembly-level constrained-random test generator. In addition, for the architectural simulation phase of the traditional post-silicon flow we used the M5 [M507] simulator for the Alpha instruction set programs and the Bochs [Boc07] architectural simulator for the x86 ISA. Test generation and simulation for both Reversi and the traditional post-silicon flow was performed on a 3.2GHz Pentium 4 machine with 2GB of memory.

3.5.1 Performance Evaluation

In the first experiment, we compared the validation performance of the traditional post-silicon system setup of Section 3.1 with the Reversi setup that we implemented. The total time for the traditional flow consisted of i) the time to create a program on the constrained-random test generator, plus ii) the time for the instruction set simulator (either M5 or Bochs) to simulate the program and compute the golden final state of the system and iii) the execution time on the silicon prototype (this can be overlaid with the simulation time). For the Reversi flow, we should include i) the Reversi test generation time and ii) the time to execute the test on the silicon prototype. Performance for the Alpha-based processor design was measured over shorter program sequences, while longer programs were tested on the x86 architecture. The results of the experiments are presented in Figures 3.15 and 3.16. The plot in Figure 3.15 shows also the performance of a typical pre-silicon logic simulator (using a behavioral Verilog model of the Alpha design) for comparison. As the graphs indicate, the Reversi-based approach flow provides a performance improvement of 19.5 times and 21.5 times over a typical post-silicon system setup for the Alpha and x86 designs, respectively. It should be noted that in addition to eliminating the architectural simulation step from the flow, Reversi is more efficient because it operates on pre-assembled blocks. In a traditional approach, on the other hand, the generator must frequently solve complex constraints to produce valid and meaningful tests. Moreover, due to the presence of branching instructions and PC-relative branches, the program generator must produce tests in assembly language and then call an assembler to convert them to machine code. Reversi, however, does not need an external assembler, since it implements internally all functions required to generate the binary code.

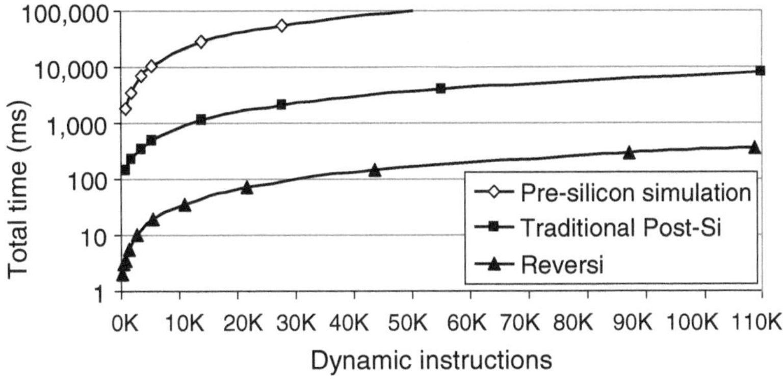

Fig. 3.15 Validation performance of a Reversi setup against a typical post-silicon system setup for an Alpha-based processor. The total time for the typical post-silicon setup includes the test program generation and the longer among simulation and test execution time. The time for Reversi includes test generation and execution. For comparison we also report the performance of a pre-silicon RTL simulator running the same tests.

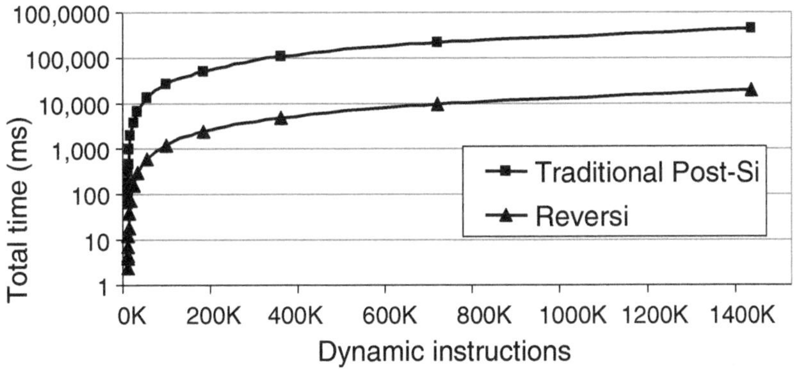

Fig. 3.16 Validation performance of a Reversi setup against a typical post-silicon system setup for an x86-based processor. The total time for the typical post-silicon setup includes the test program generation and the longer among simulation and test execution time. The time for Reversi includes test generation and execution.

3.5.2 Design Error Coverage

In the second experiment, we used an RTL implementation of a 5-stage processor pipeline running a subset of the Alpha ISA to create 20 different design variants, each containing a single bug from Table 3.3. We applied Reversi and a typical post-silicon setup to each design, and had the two frameworks generating test programs of increasing length until the bug was exposed. Note that, in order to identify a functional bug with a randomly generated program, the correct final state must first be computed by running the test on a golden model. For each buggy design, the experiment was repeated ten times using different random seeds to initialize the constrained-random test generator and Reversi. This was done to obtain statistical confidence in the results. Figure 3.17 reports the minimum, average and maximum time required for both setups to expose the bug.

Table 3.3 Functional errors injected in a 5-stage processor pipeline based on the Alpha ISA.

Bug	Description
ld_st_addr	load to store address forwarding fault
regfile_rd	faulty internal forwarding in register file read port A
fwd_mem	error in forwarding dependency resolution
fwd_reg31	forwarding through register 31 (constant 0)
ucbr_cbr	unconditional branch after conditional branch fails
fwd_wb	unnecessary forwarding from writeback stage
regfile_wr	invalid write access to register file
flush	pipeline flush on specific register file access
srl	invalid execution of logical right shift
scmp_cbr	invalid forwarding from signed compare to a branch
cbr_st	backward conditional branch after a store is not taken
ld_st_data	load to store data forwarding fault
ucmp_cbr	invalid forwarding from unsigned compare to a branch
back_cbr	specific backward conditional branch is never taken
add_over	incorrect handling of overflow on add
loop	incorrect execution of looping sequence
jsr	incorrect handling of jsr with invalid address
back_ucbr	fault in backward unconditional branch
sh_back_br	fault in branch resolution for short backward branch
ld_arith	invalid execution of a load followed by arithmetic

As the plots of Figure 3.17 indicate, Reversi can find all bugs faster than the typical post-silicon setup. Furthermore, some of the bugs, such as *loop, jsr* and *sh_back_br*, could not exposed by the post-silicon setup in any of the runs (each run could execute for at most 13,000 instructions). This is due to the unique nature

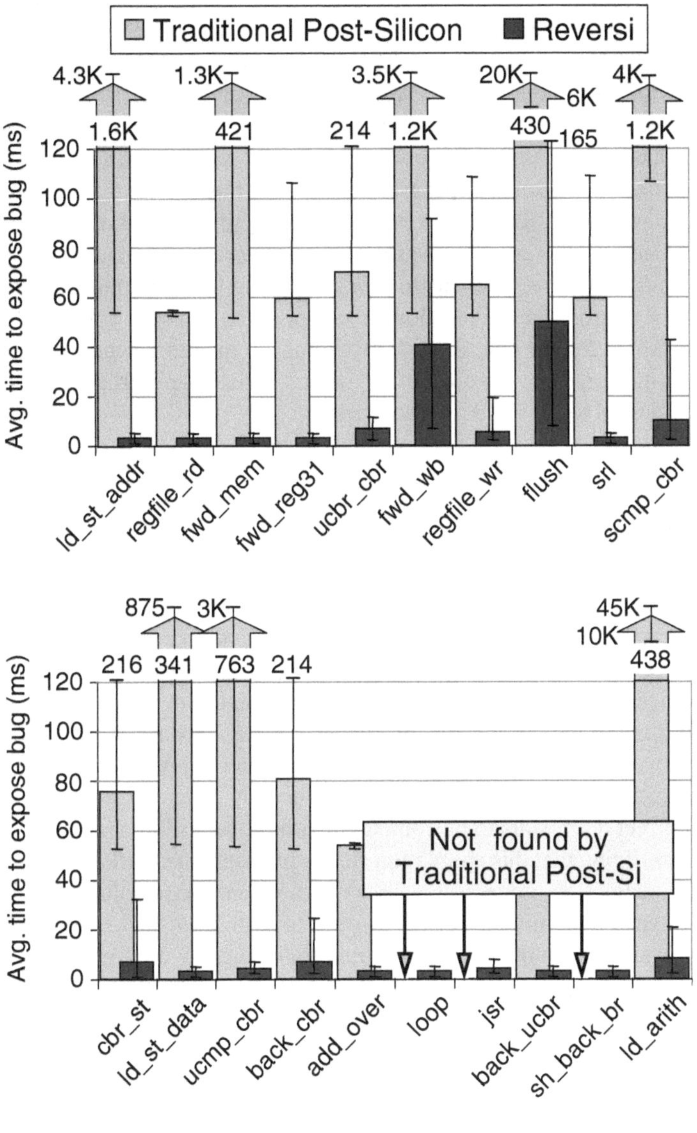

Fig. 3.17 Average time to expose functional errors in a typical post-silicon flow and Reversi. Each experiment was run ten times seeding Reversi or the constrained-random test generator with a different value each time. During each run, tests of increasing lengths were generated and executed until when the bug was exposed or until 13,000 instructions had executed. The chart reports minimum, average and maximum times to expose the bug injected in each design variant, as described in Table 3.3. Note that bugs *loop, jsr* and *sh_back_br* could not be exposed by any of the experiment runs by a typical post-silicon setup within the time boundary.

of the programs generated by Reversi – they are designed so that only correctly operating hardware produces an easily verifiable result. Thus, incorrect operations can be detected quickly at execution completion. Moreover, Reversi creates complex programs with multiple interleaved execution flows that exercise all operations in the instruction-set architecture, frequently exposing potential corner case bugs.

It's worth observing that, in several of the typical post-silicon setup experiments, such as *fw_wb*), a shorter test program exposed a bug, while a longer sequences of instructions did not. This could be explained by the random nature of the test: operations deep in the test may overwrite registers/memory locations that contain incorrect values due to earlier instructions exposing the bug, thus eliminating the evidence of the error. Therefore, a longer random program does not necessarily find more bugs than a shorter one. Reversi programs, on the other hand, are specifically designed so that any behavior corrupting the processor state is propagated to the exit point and exposed to the validation engineer.

3.6 Summary

In this chapter we overviewed traditional post-silicon validation methodologies as presented in several reports from processor vendor companies. Methodologies for post-silicon validation in the microprocessor industry can be grouped into compatibility verification, testing a processor in typical deployment settings, and system validation, stressing a broad range of features of the silicon prototype, and in particular newly developed parts of the design. This latter approach requires the development of custom system hardware and relies heavily on architectural simulation to compute correct test results to be compared with the prototype's results.

After identifying that this simulation step is precisely the bottleneck of the entire post-silicon validation process, we presented a recent methodology that leverages the performance potential of the hardware prototype itself and bypasses the slow architectural simulation phase. Test programs generated by our framework, Reversi, explore complex execution scenarios and, most importantly, produce identical initial and final architectural states in the processor, thus eliminating the need for an external component to determine if the test executed correctly. Reversi builds programs from sequences of functional blocks, which modify the state of the machine, and inverse blocks that reverse earlier operations and restore the original machine state. Individual blocks are parameterized and may consist of one or several instructions, selected at random from a block database during test generation. Reversi handles all types of ISA instructions: arithmetic (integer and floating point), logic, memory accesses, control flow and control register operations. As the experimental evaluation in Section 3.5 demonstrate, Reversi creates programs capable of finding more bugs faster than traditional constrained-random test generation techniques. Moreover, due to the omission of the architectural simulation step, Reversi can generate and execute tests approximately twenty times faster than methodologies based on a typical post-silicon system validation flow.

In the future, Reversi could be improved to require only minimal system resources to run, such as OS primitives, I/O drivers, *etc.*, so that it could execute on the same test system board as the silicon prototype under test. Test programs in this setup could be generated by a more reliable and thoroughly tested previous generation processor. This would deliver better performance than we showed in the experimental evaluation section, because of the faster communication between the Reversi generator and the prototype. We also foresee the possibility of deploying this solution in a multi-core platform, where Reversi software runs on a subset of the processor cores in the device under test, while other cores are being tested. This would allow Reversi to achieve a test generation performance that significantly exceeds the performance of today's methods and approaches validation throughput levels comparable with throughput of actual silicon.

References

[ABD+06] Miron Abramovici, Paul Bradley, Kumar Dwarakanath, Peter Levin, Gerard Memmi, and Dave Miller. A reconfigurable design-for-debug infrastructure for SoCs. In *DAC, Proceedings of the Design Automation Conference*, pages 7–12, July 2006.

[AMD05] Advanced Micro Devices, Inc. *Revision Guide for AMD Athlon™64 and AMD Opteron™Processors*, August 2005. `http://www.amd.com/us-en/assets/content_type/white_papers_and_tech_docs/25759.pdf`.

[BASS07] Tommy Bojan, Manuel A. Arreola, Eran Shlomo, and Tal Shachar. Functional coverage measurements and results in post-silicon validation of Core™2 Duo family. In *HLDVT, Proceedings of the International Workshop on High Level Design Validation and Test*, pages 145–150, November 2007.

[BEC06] Keith H. Bierman, David R. Emberson, and Liang T. Chen. *U.S. Patent no. 7133818: Method and apparatus for accelerated post-silicon testing and random number generation.* Sun Microsystems, Inc., November 2006.

[Ben01] Bob Bentley. Validating the Intel® Pentium® 4 microprocessor. In *DAC, Proceedings of the Design Automation Conference*, pages 224–228, June 2001.

[Boc07] Bochs: The open source IA-32 emulation project, September 2007. `http://bochs.sourceforge.net/`.

[CBM07] Kai-hui Chang, Valeria Bertacco, and Igor Markov. Simulation-based bug trace minimization with BMC-based refinement. *IEEE Transactions on Computer-Aided Design of Integrated Circuits and Systems*, 26(1):152–165, 2007.

[CWBM07] Kai-hui Chang, Ilya Wagner, Valeria Bertacco, and Igor Markov. Automatic error diagnosis and correction for RTL designs. In *HLDVT, Proceedings of the International Workshop on High Level Design Validation and Test*, pages 65–72, November 2007.

[EWR00] Travis Eiles, Gary Woods, and Valluri Rao. Optical probing of flip-chip-packaged microprocessors. In *ISSCC, Proceedings of the International Solid State Circuits Conference*, pages 220–221, February 2000.

[Int07a] Intel Corporation. *Intel®Core™Duo Desktop Processor E6000 and E4000 Sequence Specification Update*, November 2007. `http://download.intel.com/design/processor/specupdt/31327921.pdf`.

[Int07b] Intel Corporation. *Intel®Core™Extreme Quad-Core Processor QX6000 Sequence and Intel®Core™Quad Processor Q6000 Sequence*, November 2007.

http://download.intel.com/design/processor/specupdt/31559318.pdf.

[KAI07] Jai Kumar, Catherine Ahlschlager, and Peter Isberg. Post-silicon verification methodology on Sun's UltraSPARC T2 processor. In *HLDVT, Proceedings of the International Workshop on High Level Design Validation and Test*, page 47, November 2007.

[KDF+04] Ravishankar Kuppuswamy, Peter DesRosier, Derek Feltham, Rehan Sheikh, and Paul Thadikaran. Full hold-scan systems in microprocessors: Cost/benefit analysis. *Intel Technology Journal*, 08:63–72, February 2004.

[Lit02] Timothe Litt. Support for debugging in the Alpha 21364 microprocessor. In *ITC, Proceedings of the International Test Conference*, pages 584–589, October 2002.

[M507] The M5 simulator system, November 2007. http://www.m5sim.org.

[MDM+06] Massimiliano Melani, Francesco D'Ascoli, Corrado Marino, Luca Fanucci, Adolfo Giambastiani, Alessandro Rochhi, Marco De Marinis, and Andrea Monterastelli. An integrated flow from pre-silicon simulation to post-silicon verification. In *Research in Microelectronics and Electronics 2006, Ph.D.*, pages 205–208, June 2006.

[Rot00] Hemant Rotithor. Post-silicon validation methodology for microprocessors. *IEEE Design and Test of Computers*, 17(4):77–88, October 2000.

[SFH+03] Isic Silas, Igor Frumkin, Eilon Hazan, Ehud Mor, and Genadiy Zobin. System-level validation of the Intel® Pentium® M processor. *Intel Technology Journal*, 07:38–43, May 2003.

[WB08] Ilya Wagner and Valeria Bertacco. Reversi: Post-silicon validation system for modern microprocessors. In *ICCD, Proceedings of the International Conference on Computer Design*, pages 307–314, October 2008.

Chapter 4
POST-SILICON VERIFICATION OF MULTI-CORE PROCESSORS

Abstract. In the past four decades, microprocessor architectures have evolved from relatively simple circuits to highly optimized, performance-oriented designs with multiple parallel execution pipelines, running at gigahertz frequencies. However, in the beginning of the 21st century, designers realized that processor frequencies could no longer be increased, mainly due to power density constraints. Thus, to keep responding to users' demand for higher performance, they turned towards multi-core designs. In these architectures, multiple central processing units (cores) perform independent computations in parallel, communicating through on-chip interconnect and shared memory hierarchies. Because of the shorter and simpler pipelines and lower operation frequencies of individual cores, multi-core chips deliver much higher multi-threaded performance and demand lower power than traditional architectures. Today, with two- and four- core processor systems being mainstream, this trend continues as manufacturers roll out prototypes and application-specific systems with as many as 80 processing elements. Yet, the verification of multi-core designs has become even more complex when compared to uniprocessors, due to the concurrent interaction among cores, often leading to non-deterministic behavior. Since the majority of designs rely on shared memory protocols for inter-core communication, issues such as cache coherence and memory consistency have become dominant in the multi-core verification process. At early design stages, powerful formal tools can be wielded to ensure correctness of high-level models of these protocols. Verification of the implementations of cache coherence and memory consistency, however, still remains challenging, especially in the post-silicon domain, where much of the device's internals cannot be observed. In this chapter, we briefly overview the history and basic structure of state-of-the-art multi-core architectures and discuss the challenges of their verification. We will also overview two post-silicon solutions designed for validation of cache coherence protocols and memory consistency, setting the tone for the next chapter, where we present a comprehensive low-cost framework for the post-silicon validation of these aspects.

I. Wagner and V. Bertacco, *Post-Silicon and Runtime Verification for Modern Processors*,
DOI 10.1007/978-1-4419-8034-2_4, © Springer Science+Business Media, LLC 2011

4.1 Overview of Multi-core Processor Architectures

For about three decades, microprocessor performance and complexity has grown following the trends predicted by Moore's law, indicating that technology advances would lead to a doubling of the number of transistors on a die every 18-24 months. As individual transistors shrank, their switching speeds became faster, allowing microprocessors to run at higher frequencies and deliver higher performance. Moreover, processor designers continued to improve the architecture of their creations, devising such innovations as pipelining, super-scalar and out-of-order execution. However, in the beginning of the 21st century, the frequency race among processor manufacturers hit a wall: chips, operating at frequencies as high as 3.5GHz (Pentium 4 Prescott), became too power dense to be viable for consumer electronics. Moreover, the gains of new architectural advancements exploiting instruction-level parallelism became marginal. For example, to support higher frequencies, pipelines were stretched to 30+ stages, requiring complex branch prediction mechanisms to sustain performance. Similarly, superscalar designs could issue up to four instructions in parallel, however typical software applications rarely exhibited this much instruction-level parallelism (ILP), making wider pipelines inefficient. Yet, user demand for performance continued to increase, forcing designers to transition to a new type of processor architecture: multi-core.

Multi-core designs feature several computing elements (cores), processing data in parallel, thus executing multiple programs, or multiple threads of a single program, at the same time. Individual cores of these architectures are usually simpler than state-of-the-art uniprocessors, with shorter pipelines (about than ten stages) and less aggressive branch prediction engines. Despite that, the throughput of multi-core designs is significantly higher than that of uniprocessors for a majority of today's tasks and applications. More importantly, since multi-cores typically run at lower frequencies and can dynamically shut down unused processing elements, they require significantly less power per executed instruction than previous design generations. Finally, multi-cores provide inherent reliability through sparing: if some cores are found to be faulty, they can be disabled, leading to a reduced performance final product for a lower cost market segment (*e.g.,* Intel Core Solo [Int09], AMD Phenom X3 [AMD09]).

In the past decade, while two and four core processors have become mainstream, researchers have continued to push the envelope of architectural improvements, creating chips with performance levels never imagined before. In 2005, Sun Microsystems unveiled the UltraSPARC T1 (Niagara), which featured eight cores. Each core had only six pipeline stages, but was capable of executing four threads simultaneously, making the CPU appear to the application level as 32 independent processors. Niagara also featured a crossbar interconnect between the cores and four banks of shared L2 cache, providing more than 200 GB/s of cumulative bandwidth [KAO05]. The second generation Niagara processor (released in 2007) deepened the pipelines of individual cores to eight stages, doubled the number of threads each core could execute and boosted the floating point computation performance of the system. In the same year, Tilera introduced the Tile64 processor with 64 VLIW cores on a single die. The cores have private L1 and L2 caches and are connected by a 2D mesh network [BEA+08]. A similar 2D mesh network was used in the Intel's Polaris prototype design, which contains 80 floating-point elements and delivers a performance of 1.28TFLOPs [VHR+07]. In contrast with these systems, which have multiple identical cores, the IBM's Cell processor (released in 2006) is a *heterogeneous* multi-core design. It features a single Power Processing Element (PPE), which acts as a controller for multiple Synergistic Processing Elements (SPE), designed as application-specific SIMD-type engines [KDH+05].

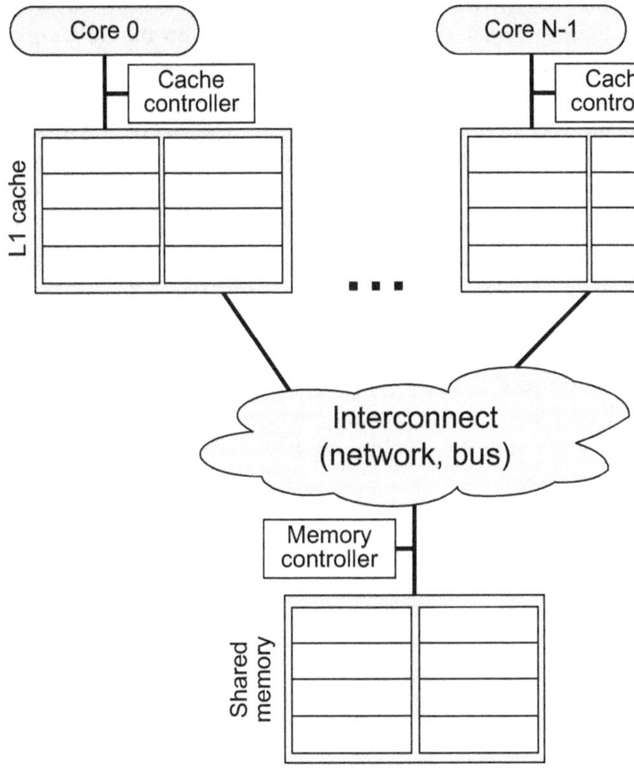

Fig. 4.1 Structure of a multi-core. Multiple cores (0 through N-1) have separate local caches, but communicate with each other and the shared main memory via interconnect. Note that data updates in a cache must eventually propagate to other nodes of the system, so to ensure proper shared memory communication. Cache coherence protocols are used to guarantee that read operations return the most recent value for each location, while memory consistency models in multi-core systems specify allowed interleaving between accesses to different addresses.

Although multi-core designs present significant performance benefits and power savings to the end-user, they exacerbate the problem of processor verification. In the majority of modern multi-core processors, the communication is implemented through shared memory. In other words, a core uses read and write operations to access locations in shared memory and distribute information, such as computational results or control messages, to other cores. To speed up memory accesses, however, designers frequently add caches to individual cores to store the most frequently accessed data and instructions. Thus, for every memory access, a local cache is checked first and, if the information is not found there, the shared memory is accessed. As a consequence, a store instruction issued by a core may simply update the value saved in its local cache, without propagating the data to the memory shared among all the cores. A typical structure of such systems is shown in Figure 4.1.

Note that in a multi-core processor there may be several cores that require access and the ability to update a same memory location, thus designers must implement mechanisms to ensure that all memory accesses return the most up-to-date value. This property is called *memory* or *cache coherence*. Typically, to guarantee coherence, designers implement a global *coherence protocol* for each of the local caches in the system. The implementation of the protocol is often distributed between cache and memory controllers, which cooperate and coordinate movements of data among local caches to ensure coherence. Another important property of any shared memory communication system is *memory consistency*, which imposes the ordering conditions on interleaving of accesses to distinct addresses during execution. Several consistency models, ranging in complexity and flexibility, have been proposed [CSG98]. For some of them consistency is enforced by the underlying hardware and is transparent to software applications, while for others ordering constraints are relaxed and users can rely on special ISA instructions to enforce the required ordering. Unfortunately, guaranteeing the correctness of cache coherence and memory consistency remains a challenging task today, even for simple multi-core processor architectures.

Pre-silicon verification of multi-core and multiprocessor systems has been a strong focus of academic and industrial communities for a long time. Since the introduction of massively parallel supercomputers, this effort mostly relied either on direct tests (programs) or constrained-random simulation techniques. Wood, *et al.* in [WGK90] used random test generation to verify cache coherence of the shared memory 36-node SPUR machine, developed at the University of California, Berkeley [HEL$^+$86]. In this work, CPU nodes of the simulated system were replaced by randomized test generators that stressed the memory subsystem and, in particular, its cache coherence protocol. The generators issued simple accesses to addresses randomly chosen from a predetermined list and then checked the results of these transactions. Since addresses on the list were shared among multiple test generators, complex interactions of the cache and memory controllers were often invoked and tested. Note that the generators were not stressing all possible coherence interactions deterministically, but were attempting to *probabilistically* validate the memory subsystem. Overall, as reported by the authors, this random testing approach was responsible for exposing more than 53% of the functional bugs found in the system before its tape-out.

Formal techniques are also recognized as a valuable tool at early stages of multi-core/multiprocessor design. For example, Murφ [DDHY92] was developed specifically for protocol specification and formal analysis. The Murφ language allows the user to specify a variety of systems, including non-deterministic protocols, through a set of declarations, rules (condition-action pairs) and invariants. Once the specification written in Murφ is compiled, it is the job of an automatically-generated verifier to identify possible invariant violations, deadlocks or erroneous states. Chen, *et al.* in [CYGC06] demonstrated how Murφ can be extended to reason about coherence even in systems with multiple levels of caches. The authors show that protocols at different levels of such hierarchy can be partially abstracted to simplify the verification process. If the abstract model exhibits violations of coherence properties, a designer may obtain a counterexample trace and then replay it on the actual protocol, to check if it is indeed a violation. This counterexample refinement is, unfortunately, a limiting factor of the approach in [CYGC06], since it cannot be fully automated and requires extensive human assistance.

Several notable approaches to multiprocessor verification adopted in industry include the work at Cray [ASL03] and at IBM [Ger03]. The former work investigated the verification of memory consistency, a significantly more challenging problem than cache coherence verification. Murφ was used in the first part of this project to create and verify an abstract formal memory model of the Cray X1 system. Formal execution traces were then extracted and re-encoded into "witness strings", stimuli that could be applied to an RTL model of the design to verify a certain property. The author estimates that this formal "guidance" allowed them to increase the overall productivity of the verification process by more than a factor of 30. Similarly, in [Ger03] Murφ was used to verify the ASCI Blue's memory protocol; however, due to the system's complexity, the author still had to resort to develop formal proofs of liveliness and deadlock absence manually, and then use these to guide Murφ in the model checking process. Yet, to validate the RTL implementation of the design, the author still had to resort to logic simulation, due to the inability of formal tools to handle such complexity.

4.2 The Challenge of Multi-core Processor Verification

Since the end of the uniprocessor era in 2000, multi-cores have become the high-performance computation engines of choice in the modern world. However, while these architectures brought unquestionable performance improvements and power consumption benefits, they added further challenges to the verification process. This is because all possible interactions between individual cores must be verified in those architectures. Thus, the heaviest burden falls on the verification of the communication protocol, rather than on datapath and pipeline control logic. As mentioned above, interaction between cores is often implemented as a shared memory protocol, with several processors communicating via an interconnect structure (bus, network, *etc.*) with each other and the main memory, as shown in Figure 4.1. Unfortunately, the latency of a memory access in such a system can be significantly higher than in a single-processor machine, since memory is physically located much further away from the cores and their local caches. A processor's request often must go through a network interface and traverse multiple hops to reach the memory controller and then return the data in response. Therefore, caches, which reside within each core/processor and amortize the access time, are vital for system's performance. On the other hand, this complicates the interaction between cores, since data values in some of the local caches may differ from those in the main memory.

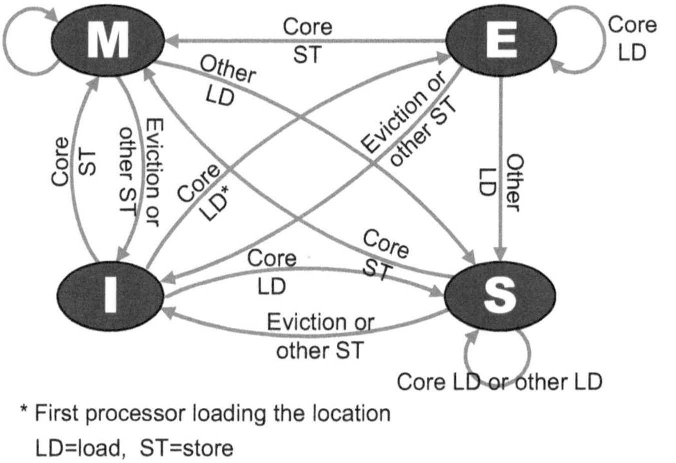

Fig. 4.2 MESI cache coherence protocol. Each memory location can be in one of the following states in each cache controller: "Modified" (M), "Exclusive" (E), "Shared" (S) or "Invalid" (I). The state is "I" when the location is not available in the cache, "E" when only the corresponding processor can modify the data, and "M" after the value has been updated in the local cache but the change is not yet reflected in other caches or main memory. The memory location is in state "S" when the corresponding data is available in the cache and, at the same time, may also be present in caches of other processors in the system.

To guarantee that all processors have a coherent view of each memory location and all data changes are propagated through the entire system with the best possible performance, a variety of *cache coherence* protocols have been proposed. As an example, in Figure 4.2 we present a finite state machine (FSM) representing the *MESI* invalidation coherence protocol [CSG98], where (from the point of view of each cache controller) a particular memory location can be in one of four states: "Modified" (M), "Exclusive" (E), "Shared" (S) or "Invalid" (I). For example, if core C1 has memory location 0x1000 in its cache in state "S", that implies that the value in C1's cache is the same as in the main memory and that same memory location can also be in the shared state in other caches.

Note that the finite state machine of the protocol shown in Figure 4.2 only reflects the view of a single processor core on the state of the memory location. Therefore, the FSM for the full system protocol for a single memory location is a *product* of *N* finite state machines, where *N* is the number of processor nodes or cores in the system. A product FSM for a MESI-based system with three cores is shown in

Figure 4.3. In this product FSM, states are labeled by combining the MESI state of each node for that particular memory location. For instance, when a location is in shared state in cores C0 and C1, while it is not cached in C2, the corresponding FSM state is "SSI". Verification of memory coherence includes determining the correctness of this system-level FSM, which encodes all possible interactions of individual nodes with respect to one memory location. The main aspects to verify in this case are absence of invalid transitions and invalid states (for example, states where several cores have the same location in "Modified" simultaneously).

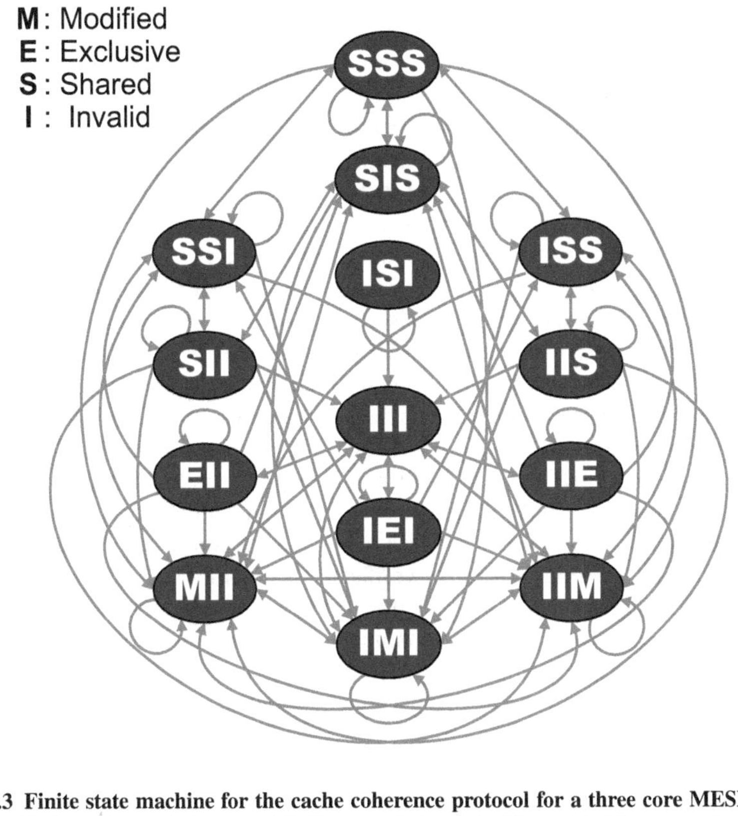

Fig. 4.3 Finite state machine for the cache coherence protocol for a three core MESI-based system. Each local core follows the MESI protocol shown in Figure 4.2. The system-level state machine is constructed as a cross-product of three local FSMs. Each cache line in the system resides in one of the states shown in the figure: for instance, a state where the first and second cores have the cache line's data in a "Shared" state and the third core does not have the data cached corresponds to the global state "SSI"

Another crucial aspect of multi-core computing is memory consistency, which defines the legal order for memory accesses in the machine. The issue of consistency arises from the fact that scalable interconnects in multiprocessors may reorder request messages, thus different cores may perceive the sequence of load and store operations in a different order. For example, the *strict consistency* model demands that all memory operations are observed in the same order in which they were issued by all processor cores. In other words, Strict Consistency requires a global arbitration mechanism among all accesses and thus, severely limits the performance of the system. Consequently, this model remains mostly theoretical and has never been implemented in a practical multi-core/multiprocessor system. On the other hand, *sequential consistency* [Lam97], requires that the same order is observed by all processors, although this may differ from the order in which accesses are actually issued. Sequential consistency can be easily implemented in systems containing a communication medium that imposes a unique ordering on all memory requests, for instance bus-based multi-core CPUs. Other models, such as *processor consistency* and *total store ordering*, force even fewer restrictions on the order in which actions may occur. However, their implementations require dedicated ISA instructions so to allow software developers to have precise control over the ordering of memory accesses, possibly at a performance cost. Memory consistency is typically hard to verify, since implemented models tend to have a wide range of distinct correct behaviors to be tested. Unfortunately, lack of thoroughness in this verification may result in escaped bugs causing unexpected software behavior.

As mentioned earlier, verification of cache coherence and memory consistency in shared memory multi-core processors is a crucial challenge faced by today's hardware designers. Even if protocols can be proven correct for a high-level abstract model by using powerful formal tools, their implementations in RTL and silicon are often significantly more complex, since they include various performance optimization features, such as translation look-aside buffers (TLBs) and reordering load/store buffers. Moreover, in RTL, individual transitions in the protocol state machine are often broken down into multiple operations, introducing numerous *implementation-level states* into cache coherence FSMs, and, thus, exponentially increasing their state space. Similarly, the problem of verification of memory consistency is exacerbated by these implementations, since the number of possible interactions between cores grows by orders of magnitudes compared to the original protocol description. Thus, protocols that are formally proven to be bug free in a high-level specification, may still be implemented with subtle, yet dangerous errors, that must be detected before system release.

The challenge of multi-core processor validation is especially pronounced in the post-silicon domain where, given a multi-core prototype, hardware engineers subject it to a variety of deterministic and randomized tests. In these tests, major errors such as deadlock, memory corruption, livelock or invalid outcome of a program, can be detected at the CPU's interface-level. However, diagnosing the root cause of such errors is a daunting task, due to severely limited visibility of the design. At this stage, the design prototype is operating at frequencies comparable to those of the final deployed product, and to access most of its internal state engineers must rely on mechanism such as scan chains, JTAG interfaces, *etc.*, which often require the execution to be suspend during state acquisition. Consequently, the timing of the events occurring within the processor may be altered, impairing the ability to reproduce the error. Furthermore, most processor's performance-related features are virtually transparent to the executing software, making it practically impossible to detect and diagnose errors in those components. For instance, a typical program does not distinguish between a load operation that hits in the cache and one that is propagated to main memory, without very accurate measurements of execution timing. Likewise, the state of cache controllers and of TLBs cannot be probed by user-level programs; thus the behavior of a coherence protocol cannot be verified without suspending execution and accessing the CPU's microarchitectural state via debug ports. However, these interruptions of the processor's execution may alter the timing of coherence and consistency events and mask hard-to-find corner case bugs, which arise due to specific timing of communication messages. Moreover, while the invariants of a cache coherence protocol are often strict and deterministic, memory consistency models frequently allow for multiple correct outcomes of an instruction sequence. Therefore, timing and interleaving of events that lead to a bug must be recorded and replayed in the same precise order to fully diagnose an issue. Special on-chip data structures, such as performance monitors and on-chip logic analyzers, increase the visibility of the silicon hardware, however, in today's competitive market, vendors often forgo these structures in favor of performance enhancements. As a result, the memory subsystem of modern multi-core CPUs frequently contains multiple critical errors. For instance, the study in [DBB08] shows that more than 10% of the errors reported in multi-core processors errata documents are related to the memory subsystem. The infamous error in the AMD's Phenom processor [Val07] is also related to the memory subsystem, in particular to the translation look-aside buffer in the shared cache. With the number of cores and complexity of cache hierarchies growing in the future, the number of such errors is bound to increase, and it will be the task of verification teams to devise novel and clever validation methodologies to discover these bugs before the system is released to the market.

In the remainder of this chapter we will overview two post-silicon techniques that enable efficient validation of shared memory communication in multi-core designs. The first solution aims specifically at verification of cache coherence protocols, minimizing the area penalty and the performance cost to the end-user of the product. The second technique is designed for post-silicon validation of memory consistency with minimal performance overhead. In order to incur a minimal performance impact, the latter solution requires the addition of several verification-specific on-die modules, and thus it comes at a noticeable area cost.

4.3 Cache Coherence Verification Using String Matching

In this section we overview CoSMa, a solution to post-silicon cache coherence validation in shared memory multi-core designs, originally presented in the work of DeOrio, *et al.* [DBB08]. The solution uses a novel distributed tracking mechanism to monitor the coherence activity in the local caches of each core, as well as in the shared second-level cache. This activity is logged and stored compactly in existing on-chip structures to minimize the area overhead; it is then checked at regular intervals for inconsistencies, which reveal coherence violations. Checking is performed through a software implementation of a string matching algorithm, from which the approach derive its name: CoSMa, Coherence String Matching. This solution is geared specifically for post-silicon validation: indeed, activity logging is transparent to normal operation, minimizing interference with the actual test in execution. Moreover, CoSMa is only active during the post-silicon validation phase of the design process and can be turned off before system release to the marker, making the solution invisible to an end-user.

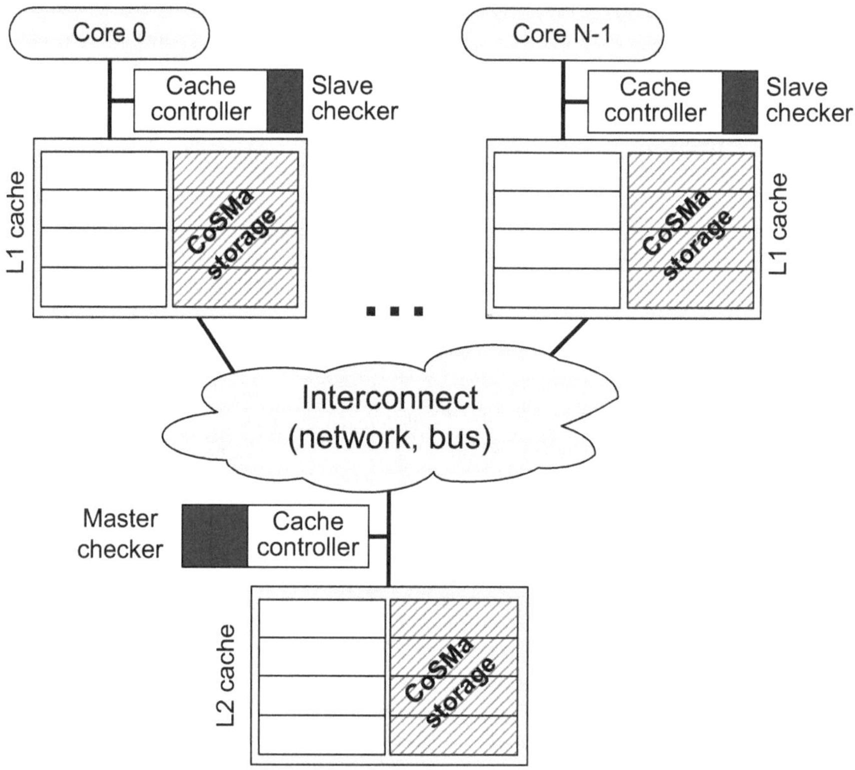

Fig. 4.4 Multi-core system reconfiguration for CoSMa post-silicon validation. A typical multi-core system with private L1 caches and a common, shared second-level cache is reconfigured during CoSMa post-silicon validation. Portions of L1 and L2 caches are converted into local and global history logs, respectively. In addition, the master and slave checker modules are enabled to periodically collect and broadcast history data. Coherence validation is implemented as a software string matching algorithm that executes on the cores of the device under validation at regular intervals. After conclusion of the post-silicon validation process, CoSMa is disabled and the history log space is reclaimed by the caches, resulting in no performance impact and near-zero area overhead in the final system delivered to the user.

An overview of CoSMa alterations to a baseline multi-core architecture are shown in Figure 4.4. As the figure shows, when CoSMa is enabled for post-silicon validation, a portion of each core's private L1 cache is converted into logged activity storage. As deterministic and/or random tests are executed by the processor, this storage dynamically keeps track of the coherence state of individual cache lines. Similarly, a portion of the second-level cache, which in this work is assumed to be inclusive, is configured to record the history of system's states associated with cache lines. The shared cache is also augmented with a master checker - a small hardware module which periodically stops and suspends the execution of the main program and interacts with slave checkers in individual cores to perform coherence validation. When the validation procedure completes, if no errors are found, the storage structures are reset and the execution of the program resumes. Note that CoSMa can be deployed in a wide range of multi-core architectures, even very different from the one in Figure 4.4, including systems with multiple levels of cache hierarchy, or different cache sharing schemes.

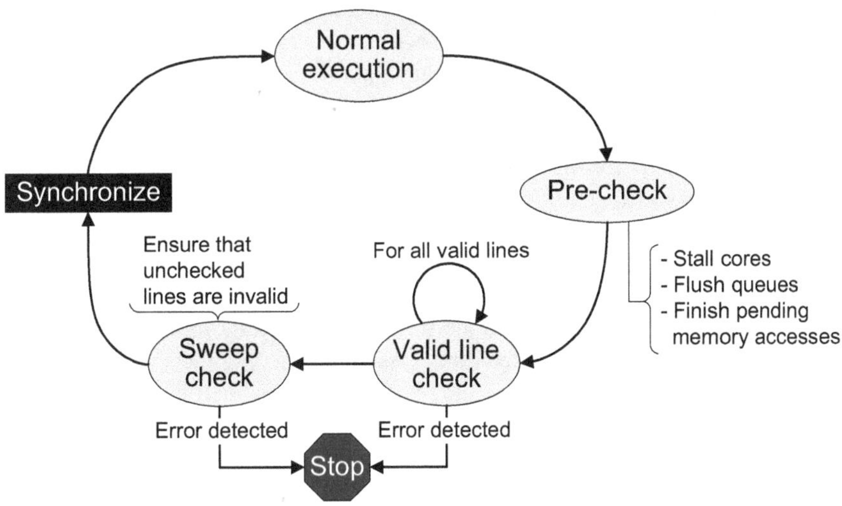

Fig. 4.5 CoSMa execution flow. When logging resources are exhausted normal execution is suspended and the system enters pre-check, where it waits for all pending accesses to complete and flushes communication hardware, such as queues, buffers, *etc.* Then the master checker begins broadcasting the global history strings of the lines in the shared cache, while checkers residing in L1 caches compare these strings to their local histories. In addition, local checkers are responsible for identification of any unchecked lines remaining in their respective caches. Upon a history mismatch or if a valid unchecked line is found, CoSMa reports an error to the user, if, however, no errors were detected, normal execution resumes after system-wide synchronization.

The master checker circuit is located within the L2 cache controller and it is invoked either with a specified time interval or when a history log in a core fills up, to conduct coherence validation. During the validation process the system behaves according to the protocol in Figure 4.5. Before validation commences, all cores suspend normal execution and all in-flight memory accesses are allowed to complete. Then the master checker starts broadcasting the global history of all the valid lines in the second-level cache, one by one. Each history string consists of the sequence of coherence states visited by a line since the last check, with additional information, such as IDs of the cores that accessed the line and timestamps of the accesses. The slave checkers receive these history strings and invoke a software-based string matching algorithm that checks the compatibility of local and global activity strings. A local and a global history string are mutually compatible if there exist at least one valid sequence of memory operations that could have generated both of them. Once all valid lines in the L2 cache have been broadcasted and checked, the slave checkers proceed to check that all remaining local lines are invalid. Finally, the system is globally synchronized, storage structures used for coherence checking are cleared, and normal program execution is resumed.

CoSMa can be configured to use one of two different coherence checking algorithms, each with different coverage/performance tradeoffs. Both are implemented in software and execute on the computational cores of the device under validation, thus minimizing area penalty. Consequently, before CoSMa is activated, the functionality of individual cores must be checked, possibly with solutions similar to Reversi (see Chapter 3). The first checking algorithm, *low-overhead*, is optimized for higher performance and minimizes the perturbation on the system under validation. In this scenario, only protocol-level states, such as "Modified", "Exclusive", "Shared" or "Invalid" for the MESI protocol described in the previous section, are saved in the line's history log. Thus, the resulting log entry consists of a string of symbols from a finite alphabet ("M", "E", "S", "I"). Once the address of the cache line and associated global history are received by a slave checker, it invokes the following three step software string matching algorithm: first, the sequence is compressed to account for cases where the L2 cache line's state remained unchanged while multiple L1 caches competed for access to the line. For instance, if two cores attempted simultaneously to modify the line several times, their histories would include subsequences such as "MIMIM", while the global history log would only contain one occurrence of the state "M". Thus, local histories with alternating valid and invalid states are compressed to a single state, ("M" for the example above). In the second step of the low-overhead algorithm, history sequences are partitioned at the invalid state boundary, since an "Invalid" state in the L1 cache may correspond to any state in the second-level cache. Note however, that a valid state in the L1 cache must be properly represented in the global history observed by L2. Thus, at the end of this step, both local and global history are comprised of an ordered series of substrings, each representing the activity occurring between two invalid states. In the last step, the algorithm matching of the substrings to a global history string. Note also that the matching must preserve substring ordering, that is, the first substring must be matched to a global history in a position before the second substring match, *etc.* Figure 4.6 illustrates this process through an example.

The authors computed the upper bound complexity of this algorithm to be $(\frac{d}{b})^{p+1}$, where d is the length of the L2 history, p is the number of partitions of L2 history, and b is the average number of states traversed in the L2 history before

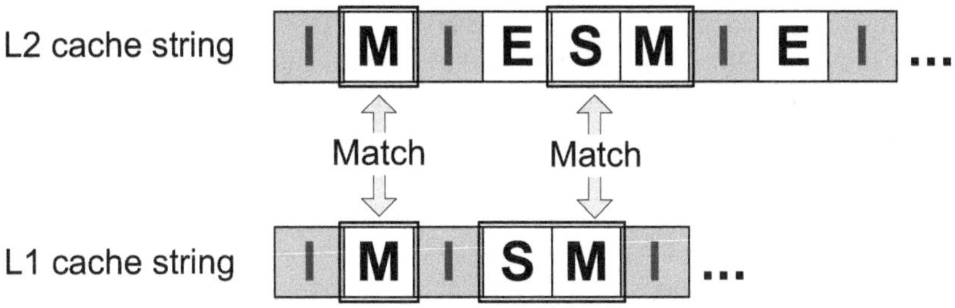

Fig. 4.6 CoSMa's low-overhead checking algorithm. Local and global history logs are first compressed, then partitioned at "Invalid" state boundary and, finally, substrings are matched for compatibility. In the example shown in the figure, the first part of a global history string includes three partitions: "M", "ESM" and "E"; while the first part of a local history comprises two partitions, "M" and "SM". Note that the two first partitions match, and the second local partition is a substring of the second global partition. Paritition ordering must be maintained in matching local and global coherence history strings.

finding each L1's substring. They also propose a coverage/performance tradeoff, which eliminates the occurrence of the worst case complexity by occasionally leaving some of the line histories unchecked. Since the need for this simplification is in practice extremely rare, the overall coverage is for the most part unaffected. It is important to remember, however, that the performance overhead of this algorithm only occurs during post-silicon validation, since CoSMa is disabled after the completion of the validation process.

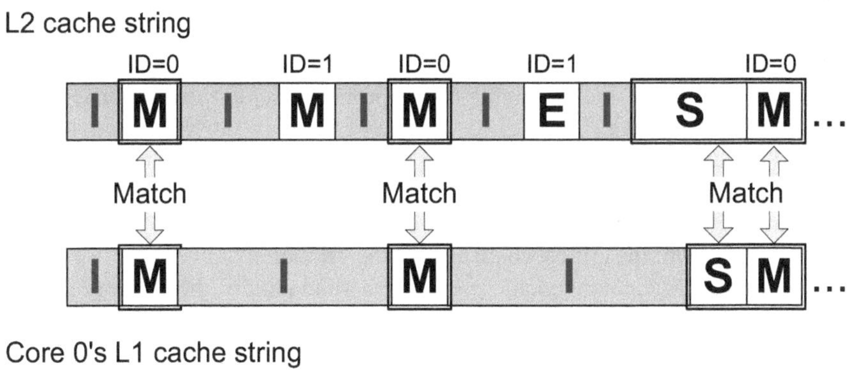

Fig. 4.7 CoSMa's high-coverage checking algorithm. In the high-coverage algorithm additional information is stored in the history log, namely the ID of the core generating the activity and a timestamp of the access. By using this information it is easy to align corresponding activity between a local core and the portion of the global activity generated by that core at the shared cache. In the example shown in the figure, the history strings have been "stretched" based on timestamp information. With this algorithm it is easy to check that both history strings observe the same (or compatible) activity during each time interval for the core under analysis.

The low-overhead algorithm keeps the history strings compact, and thus it allows for long execution intervals between the checks, since the amount of activity that can be stored in the CoSMa-reserved portion of the cache is plentiful. On the other hand, this algorithm may not identify subtle timing-related errors, which do not manifest in a corrupt coherence protocol state. Thus, the authors present an alternative algorithm, called *high-coverage*, that stores significantly more information in each log entry, namely the protocol state, a timestamp and the core ID generating the activity. During the checking phase, local and global histories are aligned based on the timestamps (Figure 4.7) and coherence compatibility is then performed. This high coverage algorithm is shown to have only linear complexity in the number of entries in the history string.

In their evaluation of the system, the authors relied on a four-node CMP system simulated with the Wisconsin GEMS simulator [MSB$^+$05]. Overall, the performance overhead, caused by periodic interruption in normal execution to perform the checks, varied from 1% to 23%, averaging to 7% for real life application benchmarks. Both checking algorithms showed similar performance overhead per check (about 1,000 cycles); however, the high-coverage algorithm was invoked more often on average. In addition, the authors evaluated the impact of seven distinct coherence bugs inspired by the Intel Core 2 Duo errata document [Int08], which they injected in their experimental platform. They noted that the low-overhead algorithm required more time to expose most of these bugs than the high-coverage solution (1M vs. 100K cycles, respectively). Moreover, the low-overhead solution was not able to detect two of the errors, which arose from timing perturbations in the system. Finally, the authors estimate that, for the OpenSPARC T1 processor, the area overhead of CoSMa checkers is below 0.002%, consisting only of the silicon area required by the logic circuits coordinating the logging activity and the checking phase.

The work of DeOrio, *et al.* presents an effective approach to post-silicon validation of cache coherence in multi-core processors, using a history string matching technique. One of the highlights of the proposed solution is the novel usage of existing on-chip data structures for storage of memory subsystem activity with minimal perturbation to system execution and negligible area overhead. The solution also allows for flexible design effort tradeoffs with two coherence checking algorithms with different performance/coverage characteristics. Altogether, CoSMa addresses several of the critical challenges in the validation of multi-core CPUs and enables high-throughput automatic post-silicon validation of the memory subsystem. However, this is limited only to cache coherence checking, falling short of the broader and more complex issue of memory consistency validation. Recall that memory consistency models describe the allowed order of interleaving between accesses to multiple addresses, thus CoSMa's tracking of coherence strings cannot provide this type of validation. In the past, it has been shown that memory consistency can be verified through analysis of the constraint graph representing the sequence of program's memory accesses and dependencies between them. However, traditionally this analysis could only be performed done offline or during short pre-silicon tests. Recently Chen, *et al.* have proposed a novel solution that brings constraint graph analysis into the post-silicon and runtime verification universe. In the next section we overview this technique and investigate its advantages, as well as its overheads. Then, in Chapter 5, we present a post-silicon memory consistency validation solution that we developed recently, which combines the principle of constraint graph analysis with the novel cache partitioning scheme proposed in CoSMa. Therefore, this framework enables efficient detection of a wide variety of memory consistency and cache coherence errors, with no performance overhead and near-zero area penalty for the end-customer.

4.4 Verification of Memory Consistency Through Constraint Graph Analysis

Memory consistency is one of the most crucial and one of the hardest properties to validate in multi-core processors. Unlike cache coherence, which describes the interactions among accesses to a single address, consistency characterizes the allowed interleaving of all memory accesses in the system. Therefore, a consistency model provides the application-level programmer with a set of invariants and rules necessary for development of multi-threaded programs. The underlying protocol enforcing these guarantees can be fairly simple and can be proven correct formally at design time. However, due to various performance enhancements, such as out-of-order execution, cache hierarchy, non-uniform memory accesses, *etc.*, the actual hardware implementation of the shared memory communication is significantly more complex. Coupled with severely limited observability at the post-silicon phase, the problem of validation of memory consistency has become a formidable challenge that cannot be resolved efficiently by traditional methods. As a consequence, researches recently started to propose solutions which augment the baseline multi-core processor with on-die detectors capable of recognizing consistency violations during the execution of post-silicon tests and benchmarks. One of such approaches has been proposed by Chen, *et al.* in [CMP08].

In that work the authors propose to detect consistency violations through the verification of the corresponding constraint graph observed during execution. A constraint graph for a segment of a program consists of vertices representing memory instructions and directed edges describing program order constraints and data dependencies between operations. These are derived and logged by individual cores during program execution and are then combined into a single graph. Memory consistency errors, which result in invalid interleaving of accesses and, thus, disagreement between the cores on the order of occurred operations, can then be identified as loops in the constraint graph. In the past, validation of consistency through graph checking has been performed offline [SS88] only after completion of the entire program. Chen, *et al.* however, rightfully acknowledge that this cannot be applied to validation of actual hardware prototypes, since it is impossible to log all of a program's memory accesses at speed and within reasonable area overhead. Instead, they developed an algorithm for slicing the constraint graph, so individual sections can be checked dynamically by dedicated hardware. To this end, each core marks instructions with unique IDs and logs the identifier of the oldest operation that was issued, but have not yet completed. When constraint subgraph construction begins, the cores merge this information and determine which instructions have retired and, thus, have completed any interaction with the rest of the system. The graph is then built only for completed instructions, which no longer can be influenced by outstanding operations. In addition to partitioning the validation problem into manageable slices, the authors also develop a graph reduction algorithm, so that only information essential to consistency verification is logged. This is performed by applying transitive closure to the data in the logs and discarding vertices that do not have inter-core dependence edges. As a result of these optimizations, program execution, graph construction and graph analysis can be pipelined and performed efficiently concurrently, while the amount of data that must be logged between checks is fairly small and does not require overwhelming storage resources.

The implementation of the solution in [CMP08] is presented in Figure 4.8, where dark blocks indicate units added to the baseline CPU design for consistency validation. In particular, each core's cache is extended to record the local history of accesses to its lines. When lines are moved between caches due to coherence activity, snapshots of these history logs are piggybacked onto the coherence traffic, so that dependency edges between instructions of different cores can be identified. These inter-core dependence edges are derived dynamically by local observer modules and are stored in their internal buffers. The authors also add a victim cache to catch lines evicted from local caches of individual cores to prevent losing the related history information. Periodically, the local observers communicate their logs over dedicated point-to-point links to a centralized hardware graph checker, which is responsible for identifying the slicing points in the logs, building the constraint graph and analyzing it for presence of loops. Construction and analysis in this framework are done in parallel with main program execution, so this approach incurs no performance overhead. A small penalty is payed to transfer history snapshots between the cores together with coherence traffic (this is estimated to require approximately 5% additional bandwidth), while the greatest impact of the approach is in terms of silicon area. In the dual core system that the authors used for evaluation, local caches had to be augmented by 30% in area to accommodate local observers and history storage. An additional penalty is incurred by the victim cache (whose size is estimated to be 3/4 of an individual L1 cache) and the central graph checker (approximately the size of an individual L1 cache). Unfortunately, often manufacturers are extremely conservative when it comes to using on-die real estate for features that do not improve the performance of the final system. Thus, the footprint of this validation collateral may be just too large to enable its adoption as a post-silicon solution. Yet, the need to validate coherence and consistency in multi-cores has become more and more important with each new generation of designs and it may be soon a critical need. In the next chapter we present a framework that provides similar validation functionality as the one just discussed, but incurs significantly lower area overhead through reuse of existing on-chip resources and implementation of the analysis algorithm in software. The tradeoff of this solution, called Dacota, is in performance overhead, since it requires temporary suspension of the execution to

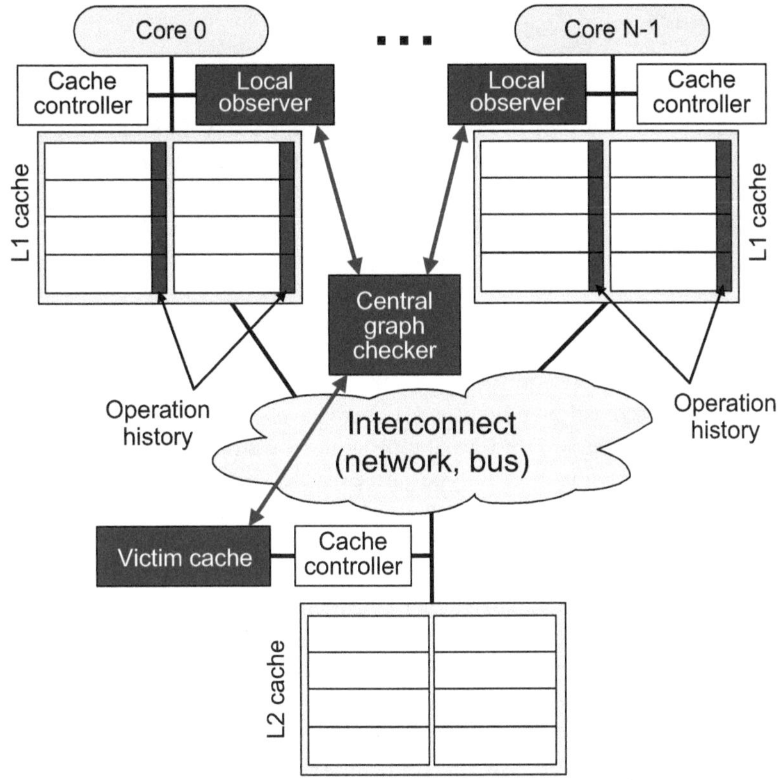

Fig. 4.8 Memory ordering validation using constraint graph checking. The baseline multi-core processor is augmented with units that observe the behavior of different memory accesses. In particular, cache lines are extended to retain information about the history of operations performed at the corresponding address, while local observer modules record inter-core dependences observed by individual cores. A small victim cache is also added to retain the information related to evicted cache lines. The logged data is periodically transferred to a hardware graph checker via dedicated point-to-point links. The graph checker constructs a dependency graph of the executed program slice concurrently with the execution of the next program block. This graph consists of vertices representing individual memory instructions and edges characterizing dependencies among them. The graph is then checked for presence of cycles, revealing memory ordering violations.

conduct the validation analysis from time to time. Note however, that this penalty is payed only during the post-silicon validation phase, so the end-user's experience is not impacted.

4.5 Summary

In this chapter we discussed the challenge of verification of multi-core processors, an increasingly popular type of processor architecture today. In multi-core designs, different computational cores execute independent instruction streams in parallel, communicating through a shared memory subsystem. To speed up the execution of memory accesses, these systems are commonly equipped with several levels of cache hierarchy. Consequently, communication among processor's cores becomes complex and often harder to verify. The two main properties of shared memory designs that must be verified to ensure correctness of operation are cache coherence, which guarantees that all load operations obtain the most recently written value, and memory consistency, which imposes the allowed order of interleaving among accesses. While verification of high-level coherence protocols and consistency models can be done efficiently with formal tools at the pre-silicon level, the actual RTL implementations and silicon prototypes are extremely hard to validate, due to the enormous state space of the global protocol state machine and to the perceived non-determinism of the memory subsystem. To aid the validation of these critical aspects, researchers have recently began to propose post-silicon solutions that reconfigure a processor to record the behavior of the memory subsystem and later check it for coherence and consistency violations. The first solution that we discussed in this chapter, CoSMa, is limited to the verification of cache coherence protocols, while the second technique is designed to detect memory consistency violations through constraint graph analysis. In the following chapter, we present Dacota, a comprehensive solution for post-silicon validation of memory subsystems, which combines the low area penalty of CoSMa with the broad validation capabilities of the constraint graph analysis technique.

References

[AMD09] AMD Phenom™X4 Quad-Core and AMD Phenom™X3 Triple-Core Processors for Home, 2009. `http://www.amd.com/us-en/Processors/ProductInformation/0,,30_118_15331_15332,00.html`.

[ASL03] Dennis Abts, Steve Scott, and David J. Lilja. So many states, so little time: Verifying memory coherence in the Cray X1. In *IPDPS, Proceedings of the International Symposium on Parallel and Distributed Processing*, page 11.2, April 2003.

[BEA+08] Shane Bell, Bruce Edwards, John Amann, Rich Conlin, Kevin Joyce, Vince Leung, John MacKay, Mike Reif, Liewei Bao, John Brown, Matthew Mattina, Chyi-Chang Miao, Carl Ramey, David Wentzlaff, Walker Anderson, Ethan Berger, Nat Fairbanks, Durlov Khan, Froilan Montenegro, Jay Stickney, and John Zook. TILE64™processor: A 64-core SoC with mesh interconnect. In *ISSCC, Proceedings of the International Solid State Circuits Conference*, pages 88–598, February 2008.

[CMP08] Kaiyu Chen, Sharad Malik, and Priyadarsan Patra. Runtime validation of memory ordering using constraint graph checking. In *HPCA, Proceedings of the International Symposium on High-Performance Computer Architecture*, pages 415–426, February 2008.

[CSG98] David Culler, Jaswinder P. Singh, and Anoop Gupta. *Parallel Computer Architecture: A Hardware/Software Approach.* Morgan Kaufmann, first edition, 1998.

[CYGC06] Xiaofang Chen, Yu Yang, Ganesh Gopalakrishnan, and Ching-Tsun Chou. Reducing verification complexity of a multicore coherence protocol using assume/guarantee. In *FMCAD, Proceedings of the Formal Methods in Computer Aided Design Conference*, pages 81–88, November 2006.

[DBB08] Andrew DeOrio, Adam Bauserman, and Valeria Bertacco. Post-silicon verification for cache coherence. In *ICCD, Proceedings of the International Conference on Computer Design*, pages 348–355, October 2008.

[DDHY92] David L. Dill, Andreas J. Drexler, Alan J. Hu, and C. Han Yang. Protocol verification as a hardware design aid. In *ICCD, Proceedings of the International Conference on Computer Design*, pages 522–525, October 1992.

[Ger03] Steven German. Formal design of cache memory protocols in IBM. *Formal Methods in System Design*, 22(2):133–141, 2003.

[HEL+86] Mark D. Hill, Susan Eggers, Jim Larus, George Taylor, Glenn Adams, B. K. Bose, Garth A. Gibson, Paul Hansen, Jon Keller, Shing Kong, Corinna Lee, Daebum Lee, Joan Pendleton, Scott Ritchie, David A. Wood, Ben Zorn, Paul Hilfinger, Dave Hodges, Randy H. Katz, John Ousterhout, and Dave Patterson. Design decisions in SPUR. *IEEE Computer*, 19(11):8–22, November 1986.

[Int08] Intel Corporation. *Intel®Core™2 Duo and Intel®Core™2 Solo Processor for Intel®Centrino®Duo Processor Technology Specification Update*, October 2008. `http://download.intel.com/design/mobile/SPECUPDT/31407918.pdf`.

[Int09] Intel Corporation. *Intel®Core™Solo processor - Intel®Core™Solo Processor support*, 2009. `http://www.intel.com/support/processors/mobile/coresolo/`.

[KAO05] Poonacha Kongetira, Kathirgamar Aingaran, and Kunle Olukotun. Niagara: a 32-way multithreaded SPARC processor. *IEEE Micro*, 25(2):21 – 29, March 2005.

[KDH+05] James A. Kahle, Michael N. Day, H. Peter Hofstee, Charles R. Johns, Thodore R. Maeurer, and David Shippy. Introduction to the Cell multiprocessor. *IBM Journal of Research and Development*, 49:589–604, 2005.

[Lam97] Leslie Lamport. How to make a correct multiprocess program execute correctly on a multiprocessor. *IEEE Transactions on Computers*, 46(7):779–782, 1997.

[MSB+05] Milo M. K. Martin, Daniel J. Sorin, Bradford M. Beckmann, Michael R. Marty, Min Xu, Alaa R. Alameldeen, Kevin E. Moore, Mark D. Hill, and David A. Wood. Multifacet's general execution-driven multiprocessor simulator (GEMS) toolset. *SIGARCH Computer Architecture News*, 33(4):92–99, 2005.

[SS88] Dennis Shasha and Marc Snir. Efficient and correct execution of parallel programs that share memory. *ACM Transactions on Programming Languages and Systems*, 10(2):282–312, 1988.

[Val07] Theo Valich. AMD delays Phenom 2.4 GHz due to TLB errata. *The Inquirer*, November 2007. `http://www.theinquirer.net/inquirer/news/995/1025995/amd-delays-phenom-ghz-due-tlb`.

[VHR+07] Sriram Vangal, Jason Howard, Gregory Ruhl, Saurabh Dighe, Howard Wilson, James Tschanz, David Finan, Priya Iyer, Arvind Singh, Tiju Jacob, Shailendra Jain, Sriram Venkataraman, Yatin Hoskote, and Nitin Borkar. An 80-tile 1.28TFLOPS network-on-chip in 65nm CMOS. In *ISSCC, Proceedings of the International Solid State Circuits Conference*, pages 5–7, February 2007.

[WGK90] David A. Wood, Garth A. Gibson, and Randy H. Katz. Verifying a multiprocessor cache controller using random test generation. *IEEE Design and Test of Computers*, 7(4):13–25, 1990.

Chapter 5
CONSISTENCY VERIFICATION USING DATA COLORING

Abstract. Today, the number of functional errors escaping design verification and released into final silicon is growing, due to the increasing complexity of microprocessor systems and the shrinking production schedules of their development. The recent, widespread adoption of multi-core processor architectures is exacerbating the problem, due to the variable communication delays of their memory subsystem, making them even more prone to subtle and devastating bugs. This deteriorating situation calls for high-efficiency, high-coverage results in functional validation, results that could be achieved by leveraging the performance of post-silicon validation, that is, all that verification activity that surrounds and applies to prototype silicon hardware. The orders-of-magnitude faster testing in post-silicon enables designers to achieve much higher coverage before market release, but only if the limitations of this technology concerning bug diagnosis and internal node observability can be overcome. In this chapter we demonstrate the full performance of post-silicon validation through Dacota, a new high-coverage solution for the validation of memory consistency models in multi-cores. When activated, Dacota reconfigures a portion of the cache storage to log the activity of memory operations using a compact data-coloring scheme. Logs are periodically aggregated and checked by a distributed software algorithm running *in-situ* on the processor cores to verify the correct ordering of the memory operations observed. When a design is ready for customer shipment, Dacota can be deactivated, releasing all the cache storage to mainstream data and instruction caching. The only remaining mark of Dacota is a small silicon area footprint, less than 0.01% of the area for an open source multi-core design used in our evaluation (three orders of magnitude smaller than previous solutions). We found in our experimental analysis that Dacota is effective in exposing a variety of memory subsystem bugs, and that it delivers high design coverage capabilities at a 26% performance slowdown for real-world applications, incurred only when Dacota is active during post-silicon validation.

I. Wagner and V. Bertacco, *Post-Silicon and Runtime Verification for Modern Processors*,
DOI 10.1007/978-1-4419-8034-2_5, © Springer Science+Business Media, LLC 2011

5.1 Introduction

The advent of multi-core processors pushed computer systems towards new levels of performance, and brought hardware manufacturers face to face with the challenge of verifying these designs, which feature complex memory subsystems, often manifesting variable data transfer delays, which in turn may cause a non-deterministic execution stream by the multi-core processor. The deteriorating situation leads to an increasing number of subtle, yet devastating, bugs slipping into production silicon, and calls for high-efficiency, high-coverage validation. To address the challenge, functional post-silicon validation, where experiments are run directly on newly manufactured prototype hardware, is becoming a vitally important aspect of today's design process. The key advantage of post-silicon validation is that its raw performance provides significantly faster testing and analysis than pre-silicon simulation, thus promising to deliver much higher coverage before the product is released to customers. On the other hand, testing a manufactured chip entails an almost complete loss of observability of the design's internal signals and state. As we described in Section 2.2, although internal values can be accessed through several debug and testing mechanisms, the delicate timing of events in the memory subsystem of multi-cores may be perturbed in the process. Alternative solutions, such as internal trace structures that sample and store the system's activity at speed and with minimal perturbation, occupy precious die area and are unlikely to be deployed in competitive commercial designs. Overall, new solutions are needed that address the validation problem in post-silicon effectively and efficiently: providing accurate replay, supporting bug diagnosis activities and with a zero, or near-zero, area footprint. We discuss, here and in the previous chapter, very recently research solutions that strive to accomplish these goals.

In the previous chapter we described CoSMa, a post-silicon cache coherence validation technique that claims a portion of the system's local and shared caches to hold activity history, and then periodically validates this activity through a software string matching algorithm. Note that CoSMa focuses only on the validation of coherency protocols and could not be applied to the verification of memory consistency, *i.e.,* the interleaving of accesses to different addresses. In that chapter we also overviewed a solution proposed by Chen, *et al.* [CMP08] that brings post-silicon validation further, enabling validation of memory consistency invariants in complex multi-core designs. Yet the area penalty incurred by this technique is quite large, making it unlikely to be deployed in commercial multi-core architectures. This chapter discusses in detail a lightweight and comprehensive technique to detect functional errors in cache coherence and memory consistency protocols in multi-cores during post-silicon validation. The solution, called Dacota (Data-coloring for COnsistency Testing and Analysis), was originally presented by DeOrio, *et al.* in [DWB09]. It offers the benefit of high validation coverage and debugging support at a very small performance impact and near-zero area overhead. Enabled only during post-silicon validation, it incorporates a simple in-hardware activity logging mechanism that observes selected events during program execution. Periodically, a software-based validation algorithm examines the logs to detect possible violations in the ordering of memory operations, indicative of an error in coherence or consistency.

Dacota shares some common traits with CoSMa, for instance, when Dacota is enabled, an activity logging mechanism located at each level one (L1) cache stores a compact encoding of memory accesses. However, in Dacota, caches are also temporarily reconfigured to include an *access vector* associated with each line: the access vectors contain a counter "color" value, incremented with each store operation to the line and used to disambiguate distinct accesses to the same cache line during execution. In addition, after each load and store is completed, the issuing CMP core logs the address and color values of the access in an *activity log*, which is also maintained in the local cache. Thus, activity logs record the history of memory accesses in program order for each individual core. When local cache storage is exhausted, activity logs are aggregated and validated by a software-based algorithm, leveraging the same processor cores for the analysis. Validation is performed by building a graph from the activity log where vertices represent accesses and edges indicate the observed ordering between them. Correct memory ordering is then checked by inspecting the graph for cycles. By leveraging existing cache storage and CPU computation resources, Dacota incurs an extremely small silicon area overhead, and it can be completely disabled upon product shipment, leading to a zero performance impact to the end-user.

5.2 Dacota Overview

Dacota comprises a small set of hardware components to be embedded in the multi-core design to be validated, which requires minimal modifications to the existing design. In addition, a software routine must be loaded in the system to coordinate the activities when checking the correctness of the memory activity collected. A schematic of Dacota's hardware architecture is shown in Figure 5.1. In the figure, we assume a generic multi-core architecture where multiple simple processing elements (cores), each with private L1 caches, are connected via an on-chip interconnect fabric to a shared L2 cache. When Dacota is enabled, the system is reconfigured so that a portion of the cache resources are reserved for Dacota's *core activity logs* (hashed blocks in the L1 caches of Figure 5.1). The fraction of the cache used by Dacota is configurable; for instance, a simple implementation could allocate half of the cache to activity log space, and the other half to normal data storage.

Processor activity is organized into *epochs*: within an epoch, program execution alternates with Dacota's checking phase. The latter can be divided into three phases, log aggregation, graph construction and policy validation. A schematic of an epoch's activities is shown in Figure 5.2. During normal program execution, Dacota monitors memory transactions in the background, by updating and transmitting access vectors along with data, and logging snapshots of these vectors. When log storage space is exhausted, program execution is suspended, data in transit is allowed to reach its destination, and the data portions of all local caches are then frozen. At

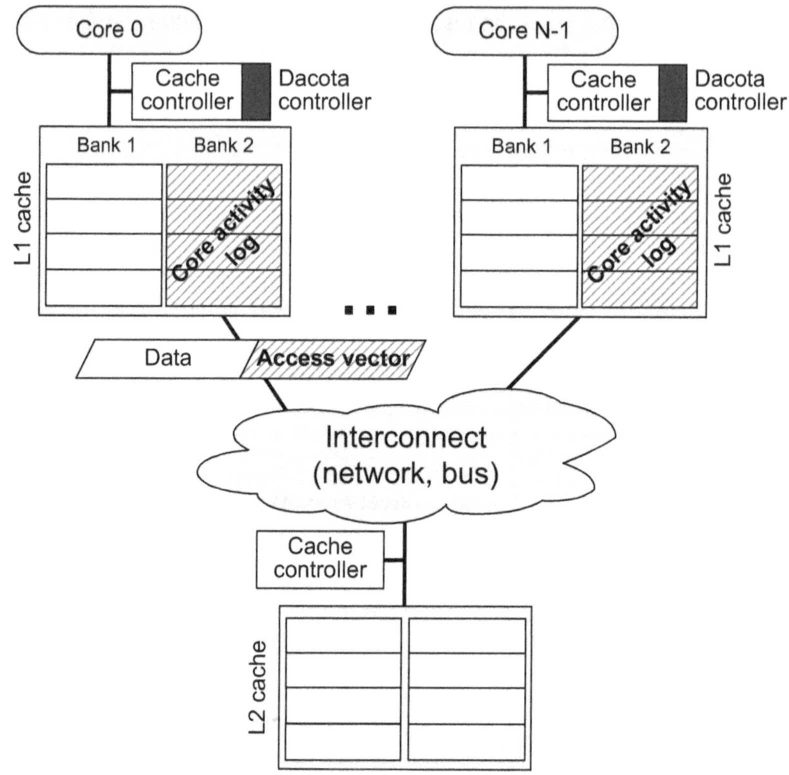

Fig. 5.1 Multi-core processor enhanced with Dacota for post-silicon validation. To augment a multi-core system to support Dacota, individual cache lines must be partitioned to include an *access vector*, which stores information necessary to track the order of memory accesses. In addition, a portion of each local cache is used as *activity log* storage, instead of normal data storage. Finally, local cache controllers are augmented to include control hardware to support Dacota's operation.

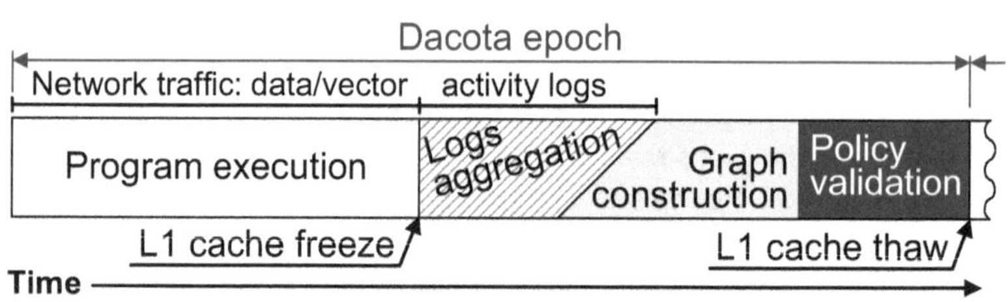

Fig. 5.2 Dacota runtime operation. When Dacota is enabled, activity is organized into epochs. Within an epoch, benchmark execution progresses normally while access vectors are transferred together with data blocks, and activity is logged in the background. After log space resources are exhausted, logs are aggregated and analyzed by a graph-based algorithm (policy validation). If an error is detected, execution is halted and the logs are presented to the engineering team for diagnosis, otherwise a new epoch begins by resuming execution.

this point, all cores drain their activity logs into a dedicated region of uncacheable memory (*log aggregation*). Next, each core in the system builds a *consistency graph* representing the ordering of memory operations. The consistency graph, which can be very easily adapted to support a number of consistency models, is examined by the *policy validation algorithm* to expose memory ordering errors. If the analysis exposes an error, information from the activity logs can be leveraged to support subsequent diagnosis and debugging. Otherwise, activity logs and access vectors are cleared and the next epoch begins.

While Dacota is active, each cache line is partitioned to include additional information alongside the data block in a form of an *access vector* that records the number and the order of store operations issued to the line. Each core modifying the data in a local cache line also updates the corresponding access vector. Traveling throughout the system with its data block, the access vector records the order of store operations to the block. A snapshot of the access vector is also stored in the local cache's *core activity log* upon each memory operation, along with the address. These logs are later analyzed to validate the ordering of memory operations. When required to support specific consistency models (such as weak consistency), Dacota may also log special memory synchronization instructions.

When logging resources are exhausted, Dacota's analysis algorithm validates the ordering of memory operations from the contents of the logs. The algorithm builds a consistency graph from the aggregated core activity logs; vertices represent memory accesses, while directed edges indicate operation sequencing as perceived by different cores. Additional edges may also be added to the graph, depending on the consistency protocol in use in the system under study. Memory ordering errors manifest as loops in the graph, thus a direct analysis of the consistency graph determines if an error has occurred with respect to the consistency model. The analysis engine can also expose coherence violations if they are manifest in the consistency graph or if they are exposed by incompatible access vectors in the activity log. Dacota's analysis algorithm is executed entirely in software on the existing CPU resources of the multi-core processor, thus it does not require any offline post-analysis nor off-chip data transfer.

5.3 Activity Logging

The activity logging system in Dacota records the order of shared memory accesses observed by different cores. To this end, Dacota maintains both an access vector attached to each cache line and activity logs residing in local caches. The information in the access vectors and logs is updated concurrently with program execution and incurs no performance overhead during this phase. However, when the log storage resources are exhausted, normal execution is suspended and the logs are aggregated and analyzed by the policy validation engine, as discussed in Section 5.4.

5.3.1 Access vector

Dacota logs the order of accesses to cache lines using a scheme based on data coloring. Each cache line is partitioned into two parts: one for program data and the other to store the *access vector*; these travel together through the interconnect and caches of the processor. Each core has a dedicated entry in the vector: the entry is updated when the corresponding core performs a store access to the cache line. The vector also includes one additional entry, reserved to count the total number of store operations to the line since the beginning of the epoch. Figure 5.3 shows several examples of access vectors, with the counter entry shown in black.

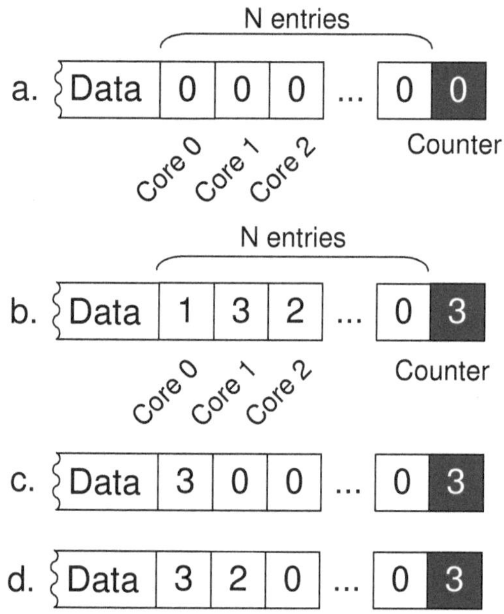

Fig. 5.3 Dacota access vector. An access vector associated with each cache line is used to track the order of store operations. **a.** Access vector at the beginning of an epoch. **b.** Access vector indicating modifications (in order) by core 0, core 2 and core 1. **c.** and **d.** Examples of aliasing due to multiple store operations by a same core. In c. it can be inferred that core 0 performed three store operations, because the other access vector entries are all 0; in d. the complete order of operations cannot be recovered: either core 0 or core 1 performed the first store operation, but we cannot infer which of the two.

At the beginning of an epoch, the counter and all entries of the vector are initialized to zero. With each issued store, Dacota automatically increments the counter and then copies the new counter value to the vector entry corresponding to the issuing core. Updates to the counter are accomplished automatically by Dacota's control hardware and do not require read-modify-write operations to be issued by the CPU. Dacota also enforces a policy that a saturated counter triggers the end of the program execution phase in an epoch: this is a necessary measure to avoid aliasing in the access vector's entries. The result of this process is an access vector with monotonically increasing counter values over time, and reflecting the chronological order in which cores modified a line.

To illustrate how access vectors operate, we show several examples in Figure 5.3.a-d. Part 5.3.a shows a vector associated with an unmodified cache line: all of its N entries and the vector's counter have a zero value. Figure 5.3.b shows the result of three store operations (as indicated by a counter value of 3), with the first issued by core 0, the second by core 2, and the last by core 1. At the beginning of an epoch, this line started with a vector as in 5.3.a; after core 0's store, the counter and the first entry were updated to 1. When core 2 issued its store operation, the counter was incremented to 2 and the value was also copied into the third entry of the vector. Finally, the latest store by core 1 changed the access vector to its present state shown in Figure 5.3.b. Observe that in this example the access vector uniquely identifies the order of all store operations to the address block. In contrast, Figure 5.3.c shows a situation of aliasing, where three store operations have occurred, yet, only the issuing core for the last one can be identified. However, for this example, the cores issuing the previous store operations can be inferred: the fact that all other entries are at 0 indicates that core 0 is indeed responsible for all the stores. In the last example of Figure 5.3.d, the unique order of the accesses to the cache line cannot be determined, since it is unknown whether core 0 or core 1 issued the first store operation. Note, however, that if we had a snapshot of this access vector before the third store operation was issued, we would be able to establish the issuing order of operations precisely.

To manage access vectors, Dacota requires the addition of a small hardware block to the cache controllers (darker blocks in Figure 5.1). This hardware component is responsible for incrementing the counter and updating the access vector's individual entries. At the cost of increased hardware complexity, it would be possible to eliminate the counter entry from the access vector and simply retrieve the counter value by extracting the highest value in the access vector entries. This alternative approach incurs higher area cost, but could be interesting for systems where storage resources for access vectors are limited.

5.3.2 Core Activity Log

While access vectors record the order of accesses to individual cache lines, *activity logs* record the order of accesses among different cache lines. Stored in a reclaimed portion of the local caches, activity logs record a series of access vector snapshots. When a core issues a load or a store operation, Dacota copies the updated access vector to the activity log, together with the type of access (load/store) and the corresponding cache line tag. The activity log is maintained as a queue and entries are allocated in program order, and then written in order of operation completion, which may differ from the program order. By leveraging the order and contents of the activity logs, Dacota can later reconstruct the order of memory operations perceived

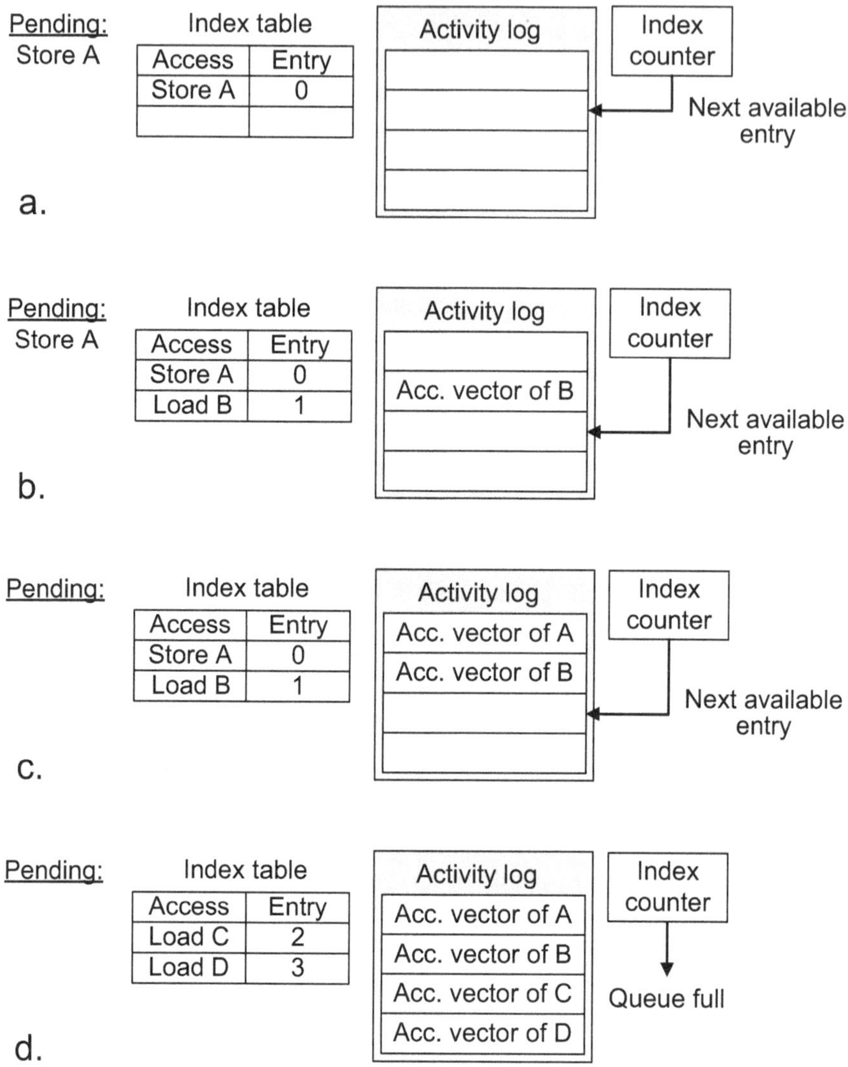

Fig. 5.4 Example of Dacota's activity log operation. The activity log records snapshots of access vectors in program order, to track the perceived order of memory operations by each local core. An index counter is used to point to the next available entry in the activity log queue. In addition, a small index table is used to connect memory accesses that have not yet completed to their corresponding log entry. In the example, **b.** the local core first issues a store operation to line *A*, correspondingly updating the index counter and index table. **b.** Then a load to line *B* is issued and completes before store *A*. **c.** Next, the store to line *A* completes and the updated vector is now logged in the first entry of the activity log. **d.** Finally, after additional load operations to *C* and *D*, the log storage is exhausted, thus triggering the checking phase of Dacota.

by each core. To reconfigure Dacota's portion of the local caches as a queue, we augment it with a simple up-counter that cycles through the allocated ways and sets. The tag array stores a portion of the location's address, while the data block stores spill-over address bits and the instantaneous value of the access vector associated with the line. In addition, since the order of completion of memory operations may be different from program order, Dacota must also maintain an index table to convert between outstanding memory accesses and entries in the activity log. Since it only needs to index outstanding memory accesses, the table can be quite small.

Figure 5.4.a-d shows an example of activity log operation. First, the core issues a store to location *A*, allocating an entry for it in the log and recording the mapping in the index table. Before the store completes, a load to address *B* is issued and completed (Figure 5.4.b), logging the access vector of line *B*. When the store completes in Figure 5.4.c, the vector of line *A* is copied to its preallocated entry. When the log fills (Figure 5.4.d), a signal is asserted and the policy validation is invoked.

In developing Dacota, we observed that logging each memory access led to prohibitively large storage requirements. Thus, we optimized the design to log a load access only if it triggered a cache miss, either because the data block is not cached, or because the copy in the cache is obsolete due to modification by another core. No loss in coverage is incurred by this optimization, when operating under the simple assumption that hit accesses to local caches are serviced correctly.

5.3.3 Activity Logging Example

An example of Dacota's logging system in operation is presented in Figure 5.5. In the example, we assume to have a CMP with two cores implementing a MESI coherence protocol and with a load/store buffer that enforces strict program order. The example shows several snapshots of a possible system execution and the corresponding logging information generated and stored by Dacota. Throughout the example, we adopt the following notation to report the information stored in the access vector: $c0|c1\,\underline{cnt}$, where $c0$ and $c1$ are the values corresponding to the entries for core 0 and core 1, and $\underline{cnt}$ is the counter's value. As an example, the access vectors in Figure 5.3.a and 5.3.d are recorded as $0|0\underline{0}$ and $3|2\underline{3}$, respectively. For simplicity we do not show the index table and the index counter of the activity log in this example.

Initially, both local (L1) caches, as well as all the activity logs, are empty (this situation is not shown in the figure). A first access operation, a store to location *A*, is issued by core 0, triggering the memory system to bring line *A* into the L1 cache in the modified state. The access vector of the cache line is then updated and stored with the cache line's data: the counter is incremented and the new value (1) is copied into core 0's entry. The resulting value $1|0\underline{1}$ and the address of *A* are also copied to the core's activity log. Figure 5.5.a shows a schematic of the system at the completion of this operation, reporting the values in *A*'s access vector and the contents of the activity logs for both cores. The letter in parenthesis in the data cache block indicates the coherence state of the cache line, in this case, "Modified".

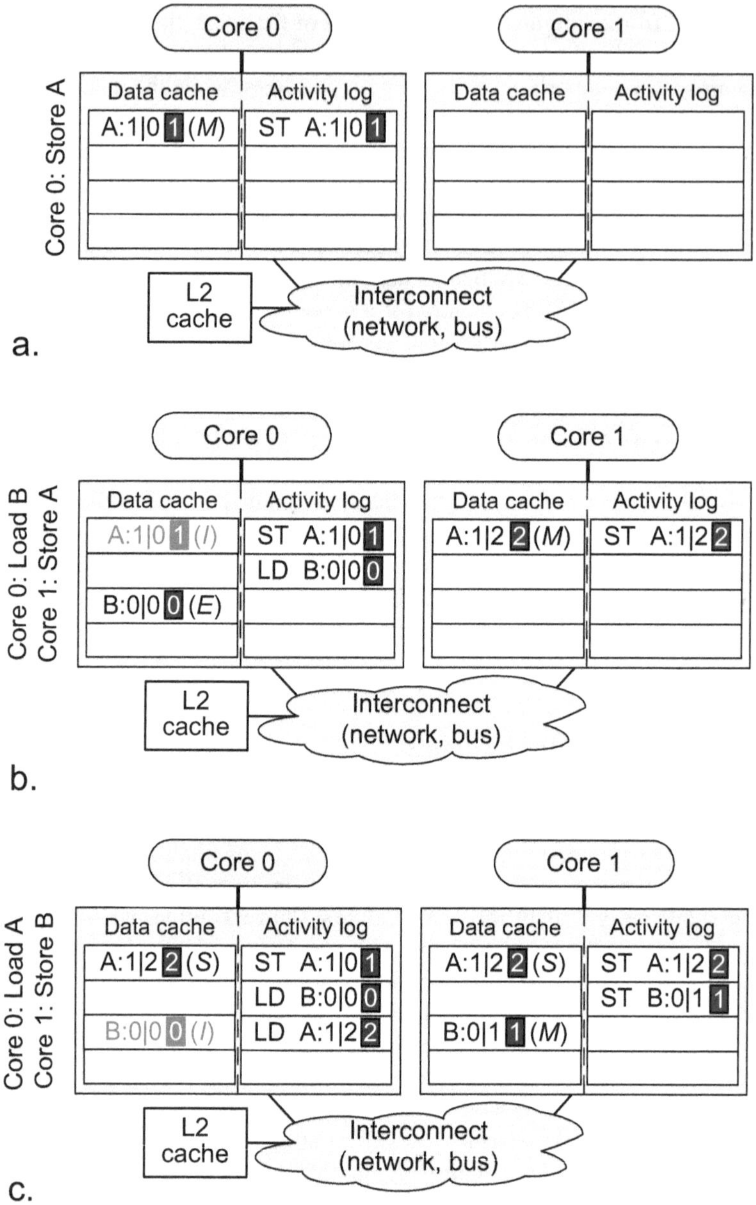

Fig. 5.5 An example of Dacota activity logging system. The example assumes a system with two cores,using a MESI coherence protocol and in-order execution of individual cores' memory operations. The three snapshots illustrates the state of the system and the information logged by Dacota after the execution of the following subsequent operations: **a.** core 0 stores to location *A*; **b.** core 0 loads *B* while core 1 stores to *A*; and **c.** core 0 loads *A* while core 1 modifies *B*.

Subsequently, core 0 issues a load to location *B* and core 1 issues a store to *A*. The status of the system at the completion of these two activities is shown in Figure 5.5.b. Note that line *B*, with access vector $0|0\underline{0}$, is brought to core 0's local cache and recorded in its activity log. Note that since this is a load operation, the access vector is not modified. The implications of core 1's store from Dacota's standpoint are as follows: line *A* must be invalidated in core 0's cache and transferred to core 1 in modified state. The vector is transferred together with the data and updated to $1|2\underline{2}$. The log of core 0 still preserves its log entry for its earlier store to *A*, while the line itself is no longer in core 0's cache (indicated in gray text in the figure). Finally, in Figure 5.5.c core 0 has just completed a load to *A* and core 1 has completed a store to *B*. The latter invalidates line *B* in core 0's cache, transfers it to core 1 and updates the vector to $0|1\underline{1}$. The former, puts line *A* in shared state. Upon loading *A* back in core 0's cache, a new entry is appended at this core's activity log.

At the end of execution Dacota uses the logs to see how the order of operations was perceived by individual cores. For instance, if the execution completed as in Figure 5.5.c, Dacota would collect the following log entries $A{:}1|0\underline{1}$, $B{:}0|0\underline{0}$, $A{:}1|2\underline{2}$ for core 0 and $A{:}1|2\underline{2}$, $B{:}0|1\underline{1}$ for core 1. From this it would be able to infer that core 0 perceived its store to *A* to occur before core 1's store to *B*.

5.4 Policy Validation Algorithm

Dacota's policy validation algorithm takes the activity logs as input, builds a directed graph representing the memory operation ordering, and finally inspects this graph to detect potential memory system's errors. The algorithm is invoked each time log resources are exhausted, either one of the activity log storage spaces become full, or one of the access vectors' counters saturates. The first step of the algorithm is to aggregate all the access logs. This process can overlap with the second phase, that is the graph construction process, which may begin as soon as the first portion of logged data is available. The graph construction process is specific to the memory consistency protocol adopted in the system under study. Finally, the last step is the policy validation proper, which is carried out by analyzing the generated graph. To minimize the area overhead incurred by Dacota, we implemented the checking algorithm in software; the algorithm is designed to execute directly on the core of the processor under validation

5.4.1 Access Log Aggregation

When a core detects that its log is full or the counter of the accessed line's vector reached the maximum value, it broadcasts a message to the global memory or second-level cache controller requesting to begin a validation phase. Upon receipt of the message, this controller requires all cores to stop execution and wait for all pending memory operations to complete. Each core, upon completion of all its pending operations, responds to the message acknowledging that it has reached a stable state. The second-level cache wait for all cores to respond, and then issues a second message requesting that all cores freeze the data portion of their L1 cache and begin transfer their local activity logs to uncacheable memory (a portion of the main memory), where the data can be accessed by all cores. Note that we use uncacheable memory to collect the activity logs so that all cores can access the information without altering the state of their local data cache, a necessary requirements to minimize the interference of Dacota with the system's mainstream execution.

5.4.2 Graph Construction

After activity logs are relocated to main memory, Dacota proceeds to build a directed graph, representing the observed ordering of memory operations. Vertices in the graph represent unique memory accesses issued during the current epoch by any core. Edges are derived from the relative ordering of events as indicated by the activity logs and from the ordering constraints specific to the consistency model adopted in the system. As graph construction progresses, Dacota also checks that coherence invariants are maintained in the system by inspecting the access vectors of individual cache lines.

```
1    Graph_Construction()
2     Graph G, Coherence_Order_Map M
3      Activity_Log L[0..N-1]
4      Foreach core c in N
5        Foreach entry e in L[c]
6           If Exists M[Address(e)]
7               Verify_Coherence(M,e)
8            Add_Coherence_Order(M,e)
9           Add_Vertex(e,G)
10          Edges E = Ordering_Edges(L[c],e)
11          Add_Edges(E,G)
12       End
13     End
```

Fig. 5.6 Pseudocode of Dacota's graph construction algorithm. The algorithm iterates through all entries of each core's activity log and generates a *coherence map*, to detect basic coherence violations (lines 6-8), and a *consistency graph*, whose edges are determined by the specific consistency model in use in the system under validation (lines 9-11).

Figure 5.6 provides a high-level pseudocode for the graph construction algorithm. The algorithm iterates over each core and log entry, building the consistency graph and, at the same time, performing a preliminary analysis to verify that all store operations to an individual cache line are compatible with a unique ordering of issuing events. In other words, it checks that for each individual cache line, all cores agree on the same order for which write operations were issued. This is a key requirement in cache coherency; Dacota verifies it by employing a data structure, the *coherence map*, that maps each line's address to an ordered list of all store operations issued for that line by any core. The algorithm inspects each entry of each activity log and, every time the line's address is encountered, the corresponding access vector is retrieved and compared or added to the coherence map (lines 6-8 in Figure 5.6). Because each access vector includes a global counter of the write operations, it is easy to order the write operations issued to a given address. In addition, since a new snapshot of the access vector is stored in the activity log every time a new write operation completes, the algorithm can determine which core issued each of the write operations. When a log entry reveals an access that was not previously observed, the ID of the corresponding core issuing the access is added to the list in the proper location. For instance, with respect to the example of Section 5.3.3 and Figure 5.5, after this analysis the coherence map would contain 0,1 for address *A* indicating that the first write operation to address *A* was performed by core 0, and the second by core 1. While scanning the activity logs it is also possible to detect and incongruence in these records. For instance, if two cores modify a cache line at the same time, due to a bug in the coherence protocol, there would be two entries in distinct activity logs, reporting the same order index for two distinct cores, and this situation could be detected in this preliminary analysis.

While building the coherence map, Dacota also uses the information in the logs to build the constraint graph for the consistency model, as indicated in lines 9-11 of Figure 5.6. For each read or write operation (to any address) occurred during the epoch a new vertex is built in the consistency graph. Thus, the algorithm builds a new vertex for each entry of each activity log: indeed, as we discussed earlier, a new activity log entry is created for each memory access by a local core. Thus a vertex is uniquely identified by (i) the core issuing the corresponding operation, (ii) the type of operation (load or store), (iii) the target address of the operation, and (iv) an index providing chronological order between operations from a same core (this can simply be the index counter value of the entry),

The edges in the graph depend on the specific consistency model adopted in the system. Below we discuss the specific edge addition rules for a number of consistency protocols. At the end, the graph is checked for loops, which are indicators of an error in the memory operation ordering, as we discuss later in this section and illustrate with examples.

Sequential consistency. The sequential consistency model imposes that all cores perceive a unique order for all load/store operations in the system. The consistency graph for this protocol includes two types of edges: *program order edges* and *address reference edges*. Program order edges are derived from the order of entries within each activity log: there must be an edge from the vertex corresponding to each entry to the vertex corresponding to the following entry in the activity log. Address reference edges connects vertices referring to a same location, of any type, from any core: each vertex has an outgoing edge to a vertex corresponding to the next operation (load or store) to that same address.

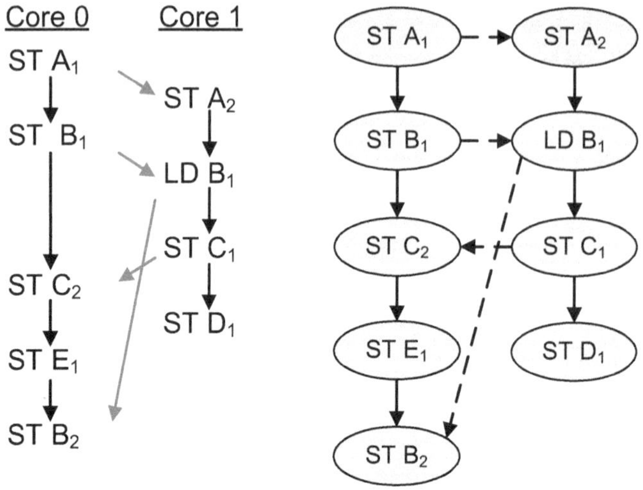

Fig. 5.7 Sequential consistency constraint graph example. The schematic on the left shows the interleaved sequence of loads and stores issued by two cores for the example. Subscripts associated with symbolic addresses distinguish among the distinct values written at those addresses by subsequent store operations; gray arrow lines connect all operations affecting a given address in chronological order. The consistency graph for this example is shown on the right: each operation corresponds to a distinct vertex; in addition solid edges represent program order constraints, while dashed ones indicate address reference dependencies.

Figure 5.7 shows an example of this construction: the left part is a schematic of the operations occurring in two cores of the processor: core 0 performed five store operations to A, B, C, E and then to B again. Core 1 performed a store to A, then loaded B and finally stored C and D. Operations by both cores are positioned so to be suggestive of their relative timing. Gray arrow lines indicate the order between operations by different cores on a same address: for instance, core 0 updated the contents of A with a store first, then core 1 modified A again. Note also that the subscript index next to the symbolic address is used to distinguish among the values written at that address by distinct store operations. All the information in this schematic can be directly derived from the activity logs: each column corresponds to the activity log of a distinct core; within a column, individual load or store operations correspond to entries in the activity log, in the same order. Subscript indexes are retrieved directly from the counter in each log entry. Gray cross-arrows are derived by comparing entries for a same address on distinct cores: the counter field and the type of operation are sufficient to determine the proper ordering among these operations. The diagram on the right in Figure 5.7 shows the corresponding consistency graph built by Dacota when operating in a machine with sequential consistency. The graph resembles closely the schematic just discussed, thus all the information necessary can be easily retrieved by inspection of the activity logs. In the graph, program order edges are indicated by continuous lines, while address reference edges are dashed.

Total store ordering (TSO). This consistency protocol relaxes the constraints of sequential consistency by allowing loads to complete before previously issued stores to a same location. When building the consistency graph for this protocol, only store operations in the activity log correspond to vertices in the graph. Edges include *program order edges* and *address reference edges*, similarly to the sequential consistency model just discussed. An additional set of edges is derived from load information: *chronological edges* represent inferred ordering among stores to different locations and can be derived from the contents of the access vectors logged due to load operations. Specifically, for each load, Dacota identifies the most recent store operation that modified that same location by inspecting the access vector, and from that it infers that any subsequent store to that same location had not been executed yet. It then retrieves the vertex corresponding to the most recent store executed executed by the same core before the load under study, and creates an edge from this vertex to the inferred *following* store to the load location. In addition, it also creates an edge from the last store that accessed the load location to the next store issued by this core. As an example, in Figure 5.8, the two chronological edges ($A_2 \rightarrow B_2$ and $B_1 \rightarrow C_1$) are derived in this way when analyzing the *LD* B_1 operation from core 1.

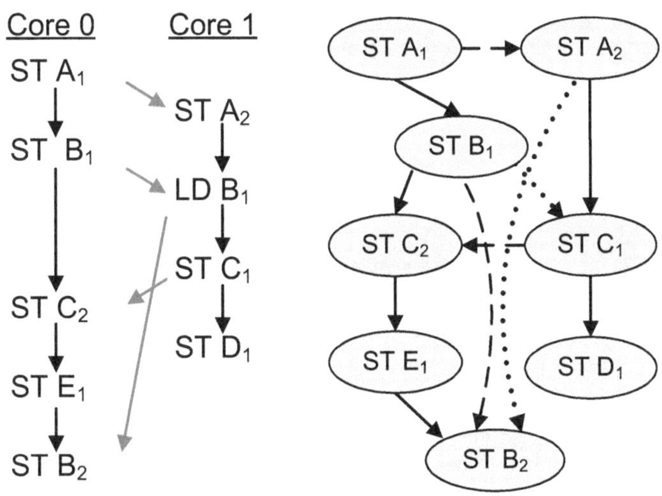

Fig. 5.8 Total store ordering consistency constraint graph example. The schematic on the left shows the interleaved sequence of loads and stores issued by two cores for the example. The consistency graph for the example is shown on the right: only store operations correspond to a vertex in the graph; in addition solid edges represent program order constraints, while dashed ones indicate address reference dependencies. Chronological edges, representing inferred chronological order derived from load operations, are shown with dotted lines.

Figure 5.8 shows an example of the consistency graph generated for total store ordering. The schematic on the left is the same as the one of Figure 5.7, and it shows the sequence of load/store operations that the two cores issued. The consistency graph on the right is derived from this sequence of operations in a machine with total store ordering. The graph contains only vertices corresponding to store instructions, as well as program order edges (continuous lines), address reference edges (dashed) and two chronological edges (dotted lines).

Processor consistency. This consistency model requires that the perceived order of store operations issued by a given core is the same throughout the entire system. This property must hold for each core; however, there is no constraint on how store operations from different cores are interleaved together, and each core may observe a different interleaving. For this model, each core must build its own consistency graph, using a process similar to that of TSO, but creating vertices only for store operations issued by the local core. Edges include program order edges, address reference edges and chronological edges derived from load operations.

Other consistency models. Weak consistency models require that only special instructions (such as memory barriers) be perceived in a unique order throughout the system, while the observed interleaving of accesses between synchronization operations may be different for different cores. In this case, beside recording loads and stores in the activity logs, Dacota also records the occurrence of these synchronization instructions and uses them as vertices in graph construction. Edges among these synchronization vertices are derived from activity log entries of ordinary memory accesses, such as loads and stores.

5.4.3 Consistency Graph Analysis

Consistency graphs in Dacota are built to reflect the order in which accesses issued by individual cores are perceived in the system. To find violations, indicators of functional errors, Dacota searches the graphs for loops, employing a modified version of the depth first search (DFS) algorithm [Knu97]. The algorithm retains the complexity of the underlying implementation of DFS, that is, $O(E)$, where E is the number of edges in the graph. In addition, to ensure maximum performance, Dacota aggressively applies transitive closure during construction, thus reducing the number of edges in the final graph.

Note that this solution does not use any additional hardware to implement policy validation: this allows Dacota to limit its silicon area overhead to a minimum, and to provide maximum flexibility in the type of analyses it can conduct. Moreover, it enables us to deploy a concurrent algorithm for the analysis, particularly for weaker consistency models, fairly common in today's multi-cores, where several distinct graphs must be constructed. For consistency policies that require the construction of a single graph, such as sequential consistency, Dacota can also leverage some concurrency, particularly in the graph analysis process where loops within the graph must be detected. Specifically, several cores will apply the same search algorithm, starting from distinct graph vertices. To further boost the performance of Dacota during policy validation, the storage previously occupied by the activity logs, which are now residing in main memory, is reconfigured to be used as regular cache space. When the analysis is completed, the Dacota portion of the cache is cleared once again to free space for the activity logs, and the data cache is thawed. A new epoch may now begin.

5.4.4 Error Detection Examples

In this section we present two examples, one of a system including a coherence error and one for a consistency error, and illustrate how Dacota identifies them.

Coherence bug example. The system in Figure 5.9 comprises two processor cores and operates with a MESI coherence protocol. At the beginning of a new epoch, core 0 loads the memory location *A* in exclusive state. This situation is depicted in part a. of the figure, where we show the address, access vector and exclusive tag for data at address *A*. In addition, Dacota adds one entry to the activity log reporting this load operation.

Subsequently, core 1 updates the value of location *A*, as shown in Figure 5.9.b. However, note that, due to a coherence protocol bug, core 0 does not invalidate *A* in its own cache, instead it changes its state to shared. When later core 0 attempts to store a new value to *A*, too, it does so by simply updating its local copy (Figure 5.9.c), since it did not invalidate that copy earlier. Now two distinct copies of this data exists on two distinct caches at the same time, a violation of the coherence protocol. As a result the corresponding entry in the activity log simply reports that core 0 performed the first store operation to *A*. This is in conflict with the activity log at core 1, which indicates that core 1 is the core that executed the first store operation to location *A*. At the end of the epoch, during the policy validation algorithm, Dacota has a chance to compare these two entries and detect and report the issue.

Consistency bug example. A more complex fault, this time in memory consistency, is shown in Figure 5.10, where we assume the same system as in the previous example, implementing a sequential consistency protocol. In this example, core 0 performs two store operations to locations *A* first, and *B* afterwards. At the same time, core 1 issues two load operations, again, to location *A* first, and *B* later. However, possibly due to an incorrect behavior of the memory controller, the load *B* operation is completed first in core 1, followed by the load of *A*. As it can be noted from the activity logs, Dacota recorded that *A* was first loaded by core 1, and later modified by core 0. After the store to *A*, core 0 also modifies *B*, which is later read by core 1.

At the end of the epoch, Dacota builds the corresponding constraint graph, shown in the lower part of the figure. Dacota can detect the sequential consistency violation dy noting the existence of a loop in the graph. The consistency bug has become evident because the two cores perceived two different ordering of load/store events, as indicated in their local activity logs.

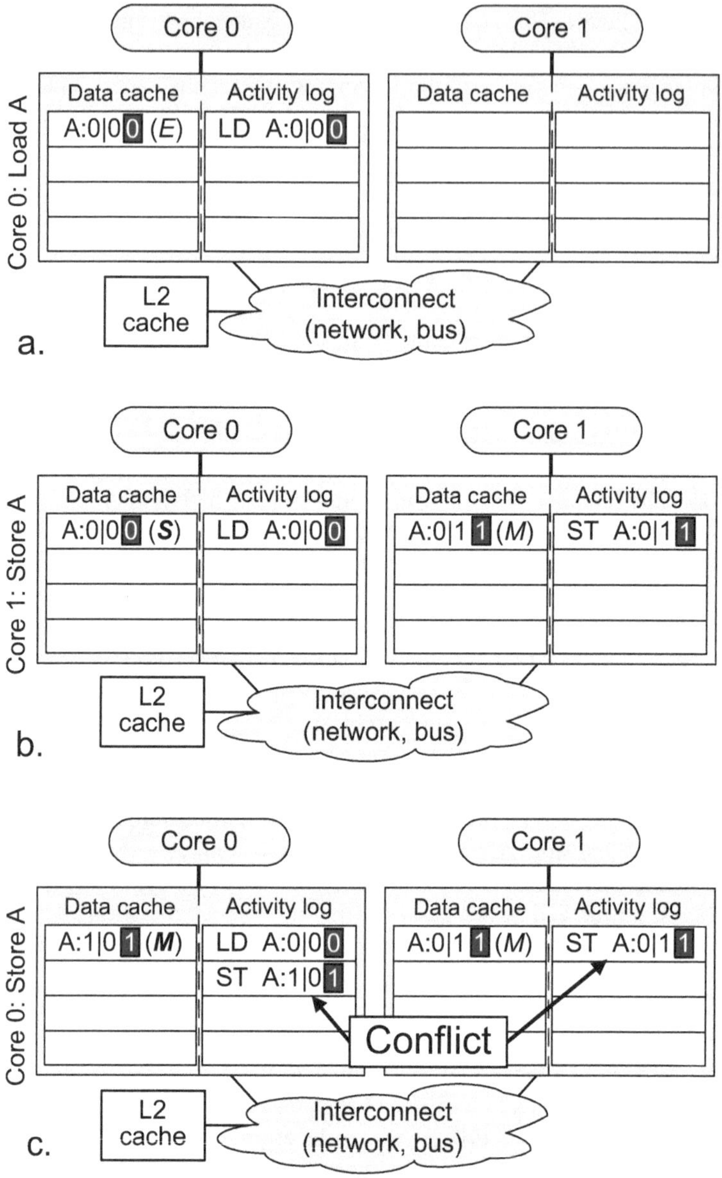

Fig. 5.9 Cache coherence error example. This two-cores system operates with a MESI coherence protocol. At the beginning of a new epoch, **a.** core 0 loads the value of location *A*, thus the corresponding cache is in exclusive state for this core. **b.** Later on, core 1 requests to modify line *A*, however, the location is erroneously updated to shared state in core 0, instead of being invalidated. **c.** When core 0's attempts a new update to cache line *A*, it updates its own local copy, leading to a coherence conflict. Because the activity logs of the two cores report that both cores where the first to update the value at *A*, Dacota can detect this error during policy validation.

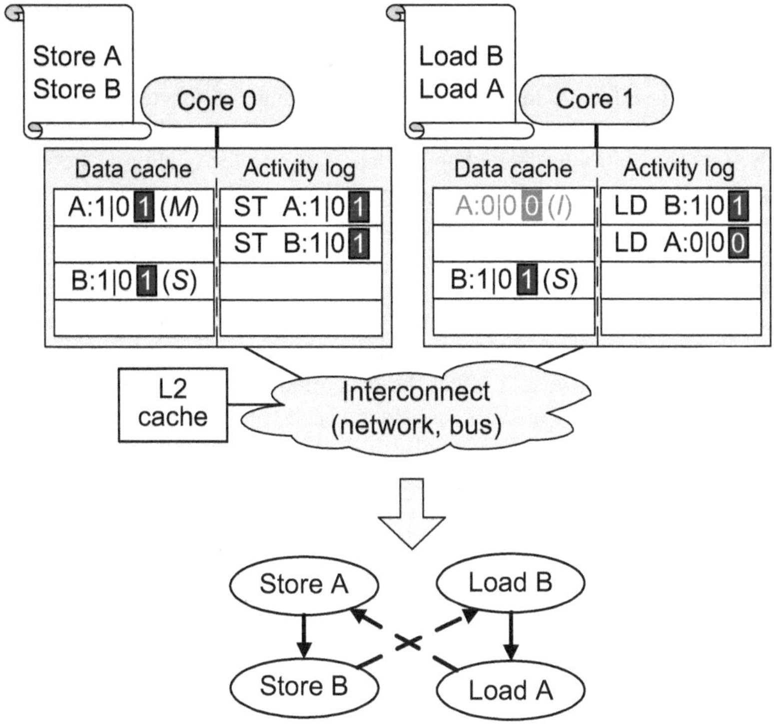

Fig. 5.10 Sequential consistency violation example. The dual-core processor operates with a sequential consistency protocol. The sequence of events in the example starts with a load *A* for core 1, followed by a store to *A* and then *B* by core 0, and finally with a load of *B* back in core 1. However, due to a bug in core 1's execution, the initial load to *A* is erroneously reordered to execute after the load to *B*. At the end of the epoch, Dacota builds the consistency graph and detects that the order in which accesses to *A* and *B* are observed by the two cores is different (by noting the existence of a loop), exposing a violation of the consistency protocol.

5.4.5 Checking Algorithm Requirements

To minimize the area overhead of Dacota, its checking algorithm is designed as software routine, instead of dedicated hardware. Therefore, for the approach to be reliable, the computational correctness of the individual cores must be determined first, as well as their ability to access main memory. Note that main memory accesses use the same subsystem that Dacota is verifying, however, bugs in the memory subsystem are unlikely to manifest errors in the analysis portion of an epoch, during graph construction and policy validation, because of the absence of memory race conditions. The cores drain the activity logs into disjoint memory regions and then use this information without overwriting it. Moreover, since the logs are aggregated in an uncacheable memory region, caches are not involved in the transfer, eliminating the chance that coherence or consistency bugs may corrupt the analysis process.

5.5 Strengths and Limitations

While Dacota is effective in exposing a diverse range of functional errors in the memory subsystem of a multi-core architecture, there are some limitations to its approach. First, its ability to detect bugs is dependent on the quality of the stimulus, that is, the program running on the system under test. Thus, Dacota can only find bugs that are uncovered by the workload running on the system. For example, workloads without shared data will not uncover coherence or consistency-related bugs, nor will very small ones that do not stress the memory system. Bugs that result in deadlock or in the system freezing are also not flagged by Dacota, as we assume that forward progress is always maintained.

Moreover, bugs that do not affect the execution's semantics evade capture by our validation system. For instance, consider a cache line shared by two cores, both in the shared state of a MESI protocol. If one processor silently (and incorrectly) modifies the line's state to exclusive the system becomes incoherent. Yet, if no store is performed to the line, the execution semantics of the program is not violated. A subsequent store operation, however, will make the error manifest, since it updates its access vector and activity logs. Furthermore, when Dacota is implemented with the optimization that activity is logged only on cache misses, as we mentioned in Section 5.3.2, errors related to the local handling cache hits would not be detected by the system. However, this aspect can be verified through properties local to an individual cache and its interface with its corresponding core, a process requiring much fewer resources than verification of system-wide cache coherence and memory consistency invariants.

5.5.1 Debugging with Dacota

One of the major strengths of Dacota is the support that it provides for the debugging effort following the detection of a bug. Indeed, the activity logs available at the end of an epoch are a useful debugging resource, providing detailed information about each cache line accessed, which cores accessed them, and when store operations occurred. With this information, an engineer can clearly see the memory activity of all processor cores, as well as system-level events leading up to the bug. Moreover, since the data portions of the cores' L1 caches are frozen during Dacota analysis, the data residing in them can also be leveraged to provide additional insights about the violation. Finally, Dacota could be augmented with small hardware components to log the state of each core's internal registers at the end of an epoch, providing engineers with more information about the program execution. These features allow Dacota to be deployed for debugging of complex and non-deterministic multi-core systems.

5.5.2 Design Considerations

In order to make Dacota a viable post-silicon solution, its design was driven by three primary goals: high coverage, low area overhead and debuggability. Performance was an additional consideration, as high execution speed is critical to our primary goal of high coverage. Early in the design phase, we considered attaching only a sequence counter to each cache line and storing these values in the activity log with each memory operation. While this minimalist setup eliminates the need for access vectors, we quickly found that the analysis algorithm corresponding to this storage mechanism was grossly inefficient. Since the sequence counter did not record the order of accesses to the line, the order had to be inferred from the logs of the other cores in the system, requiring the algorithm to walk the entire set of aggregated access logs during the construction of each vertex in the graph. With the logs placed in uncacheable memory, the performance overhead of the graph construction process outweighed the savings in log storage space. Furthermore, this scheme lost writer identification information, thus significantly hampering debuggability. With the addition of the access vector, graph construction becomes much faster, as well as it provides additional debugging information. These decisions are validated by our experimental results, which indicate that longer periods of execution between checks, as would be allowed with more compact activity logs, do not necessarily lead to better performance for the software graph analysis algorithm.

5.6 Experimental Evaluation of Dacota

The quality and efficiency of Dacota has been evaluated by simulating a multi-core system that incorporates this solution. On this platform we studied the error coverage provided, and the performance and area impact of this post-silicon solution. We also varied a number of configurations parameter to evaluate their impact on Dacota: we setup several variants of the system with different activity log lengths and analyzed their effect on consistency graph size and policy validation algorithm runtime. In addition, we measured how the amount of communication and computation overheads incurred by Dacota varies across multiple software benchmarks and activity log sizes.

The simulation framework is based on a multi-core system modeled with the Wisconsin Multifacet GEMS memory subsystem simulator [MSB+05]. The design contains 16 cores, each with a 16 entry load/store buffer, 128kB L1 cache, a single 4MB L2 cache and a 4x4 on-chip mesh interconnect. The coherence protocol adopted in the system is the MOESI directory protocol, and total store ordering is the consistency model implemented. In addition, we designed and ran simulations on a number of variants of this system: several experiments have been run on a system that used token coherence to synchronize the caches, and we also evaluated two alternative interconnect types, crossbar and switch-based. The components of Dacota are implemented as a simulator plug-in, which include methods to manipulate the access vector and manage the cores' activity logs. Finally, the policy validation algorithm is implemented in software using the Boost graph library (BGL) [SLL02]. The algorithm's runtime was calculated by simulating its execution using the SimpleScalar architectural simulator [BA08].

The set of workloads used to evaluate Dacota are a combination of real-world programs and random stimulus. We used the ten SPLASH2 benchmarks [WOT+95], compiled for a 16-cores shared memory architecture, considering execution traces of 10,000,000 dynamic instructions generated by the Virtutech Simics simulator [MCE+02]. In addition, to induce more stress on the memory subsystem, we created eight synthetic tests of directed-random stimulus with varying degrees of data sharing, each containing 1,000,000 memory accesses. Three of these benchmarks used a fairly small address space that could fit into the cores' L1 caches without eviction, while the other five used significantly larger memory ranges and could not be fully contained in the L1 caches. Additionally, we created benchmarks using the GEMS built-in random test generator executing the "barrier" and "locks" patterns.

5.6.1 Design Error Coverage

For the first experiment we created eight different variants of our design, each including a distinct coherence and/or consistency-related error. The functional errors were inspired by known issues in commercial multi-core designs reported in publicly available errata documents. We then ran our SPLASH2 and synthetic benchmarks with Dacota enabled, recording the number of execution cycles required to expose the bug. The results of this study are reported in Table 5.1, where we provide a short description of each bug. We also computed the average number of cycles that exposing benchmarks take to reach the bug (not all programs encountered all errors). It can be noted from the table that all the injected bugs were discovered by Dacota, and that all but one required less then three millions execution cycles on average, an amount that corresponding to a very small runtime in a hardware prototype.

5.6.2 Performance Evaluation

We also investigated the computation and communication overhead of running Dacota relative to the execution time of the testbench without Dacota. In this study, it was assumed that one half of the L1 cache, which we sized at 64kB, was devoted to data storage and associated access vectors, and the activity log space was limited to 256 entries fitting in 32kB. Figure 5.11 includes two plots, the top plot reports results on the SPLASH2 benchmarks, the bottom one refers to the synthetic benchmarks, that is, randomly generated programs with various degrees of data sharing and the "barrier" and "locks" tests native of GEMS simulator. The left-side vertical axis indicate the fraction of overhead with respect to a same benchmark running without Dacota. The gray and black bars in the plot refer to this axis, and

Table 5.1 Functional error coverage: number of execution cycles required for Dacota to detect each bug.

Bug name	Description of the error	Avg. cycles to expose bug
shared_store	store to a shared line may not invalidate other caches	0.252M
invisible_store	store message may not reach all cores	1.32M
store_alloc_1	store allocation in any core may not occur properly	1.93M
store_alloc_2	store allocation in a single core may not occur properly	2.27M
reorder_1	invalid store reordering (all cores)	1.38M
reorder_2	invalid store reordering (by a single core)	2.82M
reorder_3	invalid store reordering (to a single address)	2.87M
reorder_4	invalid store reordering (to a single address by a single core)	5.61M

they show the computation and communication overheads, respectively. Computation overhead accounts for the time spent running the constraint graph construction algorithm and the policy validation algorithm. Communication overhead refers to the additional time spent transferring the activity logs to uncacheable memory at the beginning of the checking phase in Dacota and then distributing the logs to individual cores for graph analysis. The dashed lines mark average values over all the benchmarks in each plot for both of these overheads. Finally we also plot in the graph the amount of data traffic imposed by Dacota to the system during normal program execution. This extra traffic, measured in kilobytes, is required to transfer access vectors along with each data line that is moved to and from local caches. The values for each benchmarks are indicated by the black stars with reference to the right side vertical axis.

Note from the figure that the average overhead for SPLASH2 benchmarks is 26%, while for randomly generated benchmarks is 150%. Overall, these values are well within the acceptable range for post-silicon techniques; indeed, overheads of 300-500% are perfectly acceptable for solutions currently adopted in this domain. The overhead for random benchmarks is somewhat higher due to the intrinsic nature of these tests, designed specifically to stress the memory subsystem. Consequently, when running these tests, the fraction of program instructions issuing memory operations is markedly higher, generating many more entries in the activity logs, which in turn must be analyzed by Dacota more frequently.

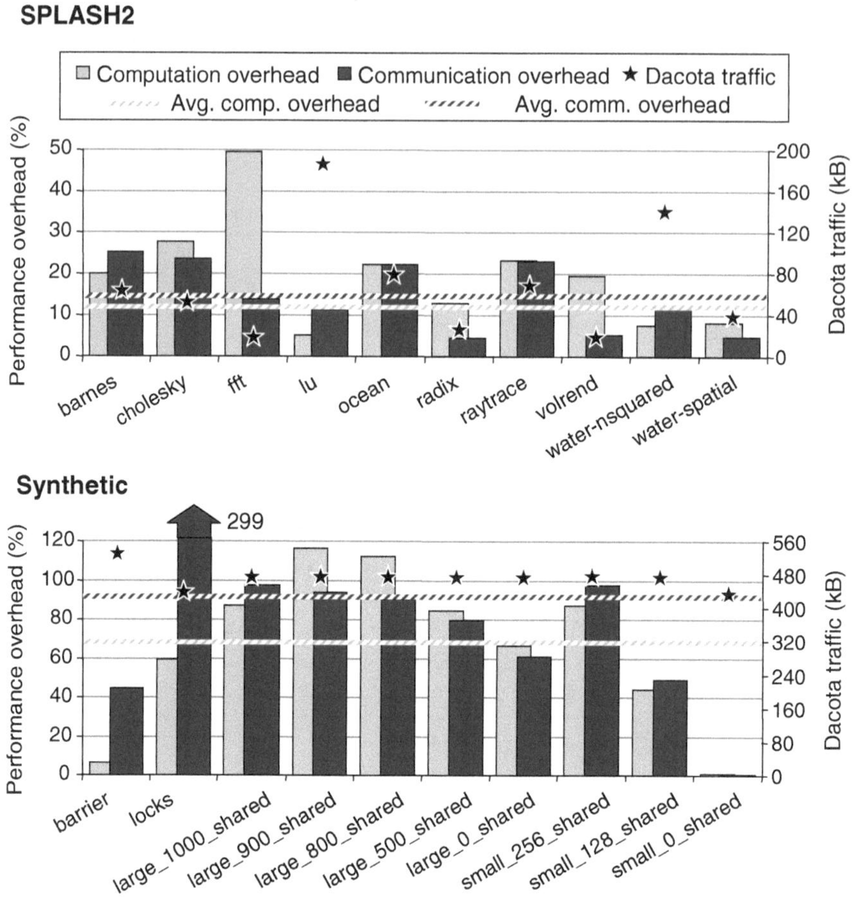

Fig. 5.11 Dacota's performance overhead. The graphs plot the performance overhead incurred by the system when Dacota is active in post-silicon validation. The gray bars report the computation overhead incurred for each benchmark and the black bars plot the communication overhead. Both bars are with reference to the left side vertical axis. The amount of additional traffic incurred during normal operation is also plotted with reference to the right side vertical axis. The top plot reports results for the SPLASH2 benchmarks, the bottom one for the synthetic benchmarks. Dashed lines indicate average values of the overheads for the set of benchmarks in the plot.

The Dacota implementation that we used in the experimental setup serializes the two phases of activity log transfer and graph construction; in practice there could be a significant overlap, since the construction of the constraint graph may begin as soon as there is at least one activity log available in uncacheable memory. If we had this optimization in place, the aggregate overhead would be smaller than that reported in Figure 5.11. It is also important to note that, since Dacota can (and should) be disabled upon product shipment, these performance overheads are only incurred during in-house validation of silicon prototypes.

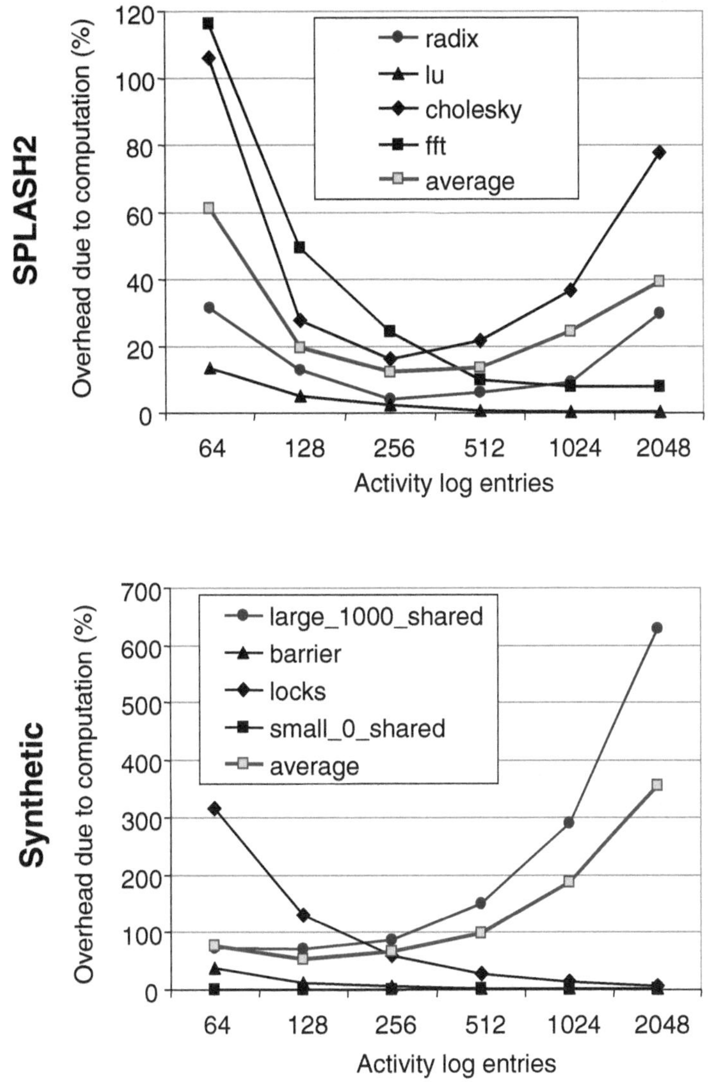

Fig. 5.12 Computation overhead vs. activity log size. The plots report the computation overhead incurred by a number of testbenches when using a varying amount of activity log storage space. The size of the activity log used in each evaluation is represented on the horizonal axis. The top part plots SPLASH2 benchmarks results, the bottom part focuses on synthetic benchmarks. Also reported with light gray markers is the average result over all SPLASH2 and all synthetic tests.

The next study focus on the impact of the activity logs storage space. For this experiment, we repeated a same simulation multiple times, each time varying the size this storage and measured the communication and computation overheads incurred. The results of the analysis are presented in Figure 5.12 and 5.13. Both figures

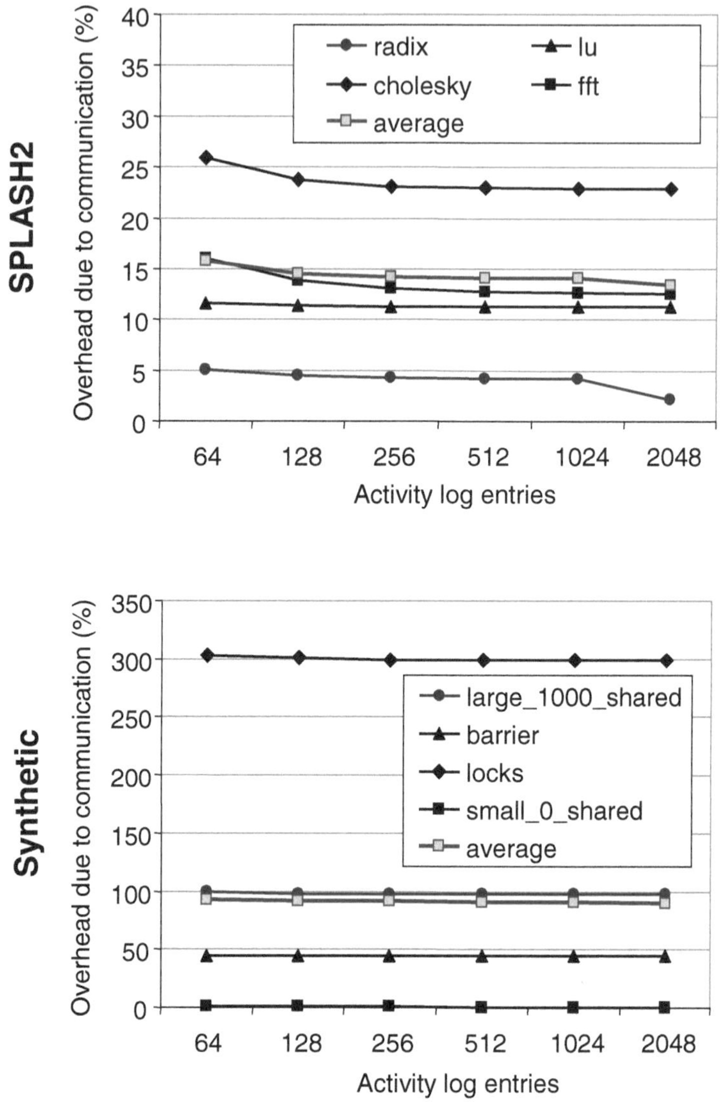

Fig. 5.13 Communication overhead vs. activity log size. The plots report the communication overhead incurred by a number of testbenches, named in the legend, when using a varying amount of activity log storage space. The top part plots SPLASH2 benchmarks, while the bottom part focuses on synthetic benchmarks. Also reported with light gray markers is the average result over all testbenches in each family: SPLASH2 and synthetic.

include two plots each, the top one focusing on SPLASH2 benchmarks, and the bottom one on the synthetic tests ("locks", "barrier" and randomly generated): since the impact of Dacota appears to be very different for these two classes of tests, we strove to conduct separate analyses throughout the evaluation. Figure 5.12 reports results with respect to computation overhead, while Figure 5.13 focuses on communication overhead. In each of the plots, the horizontal axis indicates the size, in number of entries, of the activity log storage in each local cache. The legend reports the names of the individual testbenches analyzed. The plot line with light gray markers shows the average over all tests of each family, that is, the top graph includes the average over all then SPLASH2 benchmarks we evaluated, and the bottom graph indicates the average over all ten synthetic benchmarks.

The computation overhead analysis reveals some interesting trends: for some benchmarks, such as *fft*, *lu* and *locks*, the overhead monotonically decreases with larger log storage space. However, several other benchmarks, including *cholesky*, *radix*, *etc.*, present a local minimum at medium log sizes. Most of the synthetic benchmarks, on the other hand, seem to impose more overhead when they have larger storage available for activity logs. We believe that these trends may be explained by the memory access patterns of the individual benchmarks: testbenches with high density memory accesses and with much data sharing fill quickly any activity storage and the amount of data to process in graph construction and policy validation increases both in size (because of the larger storage) and complexity (because of the high fraction of data sharing), leading to a marked increase in computation demands. Testbenches with relatively infrequent memory accesses and little sharing instead pose little challenge to the policy validation algorithm. In addition, it is not infrequent for these testbenches to trigger a checking phase because of the saturation of a counter in the access vector, rather than because of exhaustion of log storage space.

In Figure 5.13, it can be noted that the communication overhead is practically constant for all testbenches, that is, the time spent transferring activity log data is a constant fraction of the execution time. This seems a logical result since large activity log storage space requires more transferring time, but also allows for longer epochs with longer intervals of time spent in normal computation. Aggregating all our results, we found that 256 entries per activity log seems to strike a good balance leading to a generally low overhead.

We also studied the impact of different interconnect topologies on Dacota performance and overheads imposed. To this end we conducted additional experiments using the same set of testbenches and varying the size of activity log storage but over two alternate systems: one using a crossbar interconnect and one using a hierarchical switch interconnect architecture. For each of these architectures, we also evaluated the system when the coherence protocol was token coherence instead of MOESI. Overall we found that Dacota's performance was not affected in any appreciable amount by the interconnect topology; however the use of the token coherence protocol led to an 8% reduction in communication overhead. Investigating this result, we found that the ratio of traffic due to Dacota moving activity log data over normal system traffic is lower with token coherency because of the larger number of control messages present only during normal execution and required to implement this coherence scheme.

Finally, we evaluated how often Dacota must be invoked to aggregate log data and run the policy validation algorithm. This is directly related to the average length

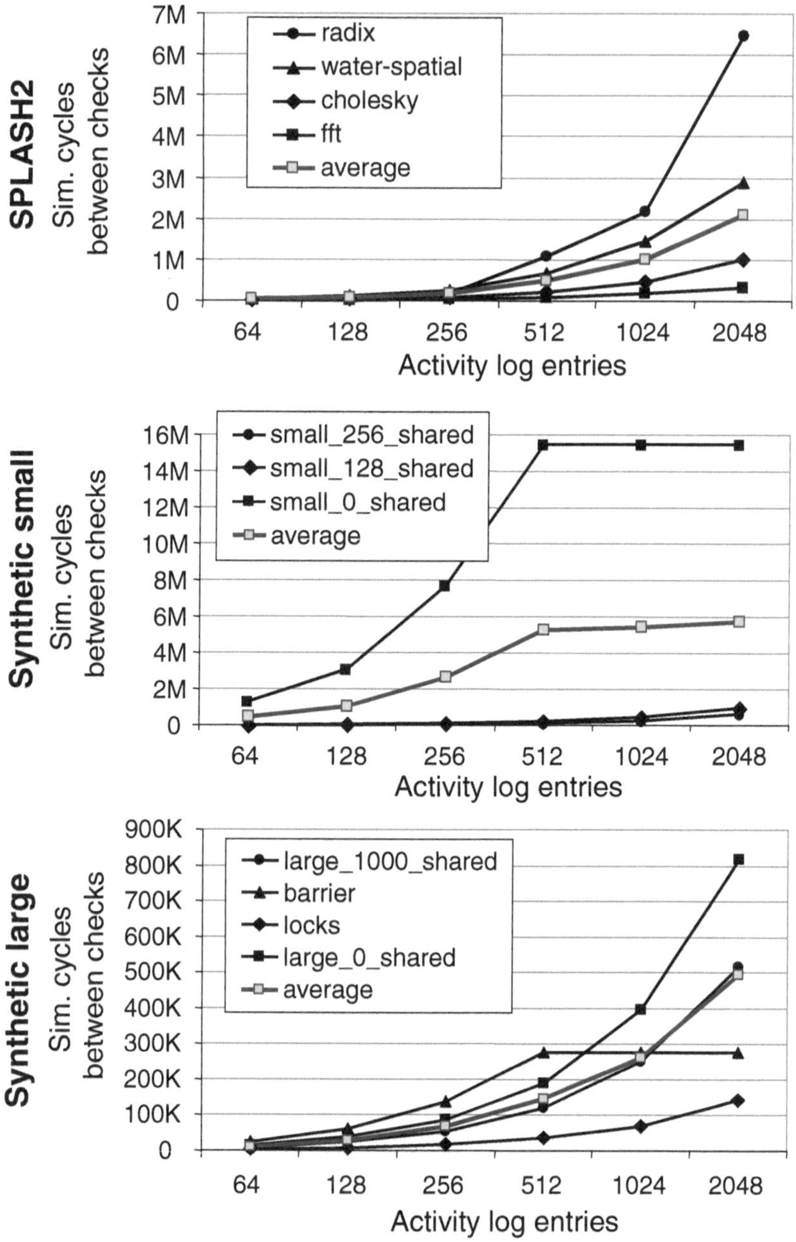

Fig. 5.14 Normal execution cycles within an epoch. The plots report the average number of cycles spent in normal program execution between each pair of checking phases, for a number of SPLASH2 and synthetic benchmarks. The graph line with light gray marker indicates the average over all SPLASH2 benchmarks (top plot), over the smaller synthetic benchmarks in the plot (middle), and over all remaining synthetic benchmarks (bottom).

of an epoch of execution. Figure 5.14 reports the results of this study indicating how many clock cycles of program execution are completed, on average, between each pair of checking phases. The top plot reports representative results for a number of SPLASH2 benchmarks, while the two bottom plots report results for the synthetic benchmarks, the smaller ones in the middle plot and the larger ones in the bottom plot. In each plot the average number of cycles between checks is graphed against the size of the activity log storage space. The plot line with light gray markers reports average information: in the top plot the average is over all SPLASH2 benchmarks, in the middle plot it is over only the testbenches in the graph, and in the bottom plot it is over all other synthetic benchmarks.

As it can be noted, and as intuition suggests, increasing log storage leads to longer intervals between checks, since more memory accesses can be recorded before the storage space saturates. Note, however, that there are a number of exception to this rule of thumb: for instance, the *barrier* benchmark does not seem to be benefiting from log storage with more than 512 entries. The plateau in this benchmark occurs because the check is triggered by access vector counter saturation, not due to the filling of the log. Benchmark *small_0_shared* has a similar pattern as well. Recall that the smaller synthetic programs in this experiment, including *small_0_shared*, have address spaces that fit completely in the L1 caches of the individual cores. In addition, this particular benchmark has no shared data, thus lines cached by the cores are never invalidated or evicted. As a result, activity logs with 512 entries of more never fill throughout the entire execution of the benchmark, and Dacota analysis is invoked only once, after the benchmark's completion.

5.6.3 Area Evaluation

In order to deploy Dacota in a multi-core system, a few small hardware components must be added. To perform a coarse evaluation of the area occupied by these components, we designed them using a register-transfer level language (Verilog HDL), synthesized them and gathered their size from the synthesis tool final report. The Dacota modules included a control block for updating the counter and access vector, a state machine to manage the activity log space, and an index table to maintain the conversion between program order and completion order of memory accesses, as discussed in Section 5.3. The modules were synthesized with Synopsys Design Compiler targeting a TSMC library at the 90nm technology node. The collective area required to deploy all these modules was reported at 5,216μm^2. For comparison, an OpenSPARC T1 [LTS$^+$07] processor occupies approximately 378mm^2. Placing a full set of Dacota modules next to each of the eight cores in this design would result in an area overhead of 0.01%, an area impact fairly negligible, even for such area-sensitive platforms as modern microprocessors. The underlying reason why Dacota can be deployed with such small area overhead lies in its ability to reuse reuse of existing hardware structures for most of its activity, such as cache storage, and its leveraging software algorithm for policy validation, instead of dedicating silicon area to implement this task.

In order to put Dacota's area overhead in perspective, we also compared its storage requirements with two other memory consistency validation solutions: the technique by [CMP08] overviewed in Section 4.4 and a *runtime* approach proposed by Meixner, *et al.* in [MS06]. We will discuss this latter approach in more details in Section 8.1. A conservative estimation of the added storage requirements for the first technique results in 171kB for a 16-node system, similar to the one used in our experimental evaluation. This includes all the overhead components in [CMP08]: extension of individual cache lines, area occupied by local observers, victim cache and central graph checker. In the case of the solution by Meixner *et al.*, at least 486kB of storage are required by the detection hardware itself, while another 432kB would be needed to implement SafetyNet, a checkpointing framework that enables runtime error recovery [SMHW02], totalling 918kB. In contrast, since Dacota is capable of reusing existing storage space for its activity logging, it requires very little additional hardware, a 2B counter and a 32B table at each node, totaling 544B in the 16-processors system used in our evaluation. Thus, considering these conservative estimates, our solution is approximately 321 times smaller than the post-silicon validation technique in [CMP08] and 1,728 times smaller than the solution in [MS06].

5.7 Summary

In this chapter we presented Dacota, a novel solution for high-coverage post-silicon validation of memory ordering in multi-core processor systems. When enabled by the verification team, Dacota stores information about the sequence of executed memory operations, periodically aggregating this data to perform a software-based policy validation. The validation algorithm is implemented purely in software to minimize this solution's area impact and executes the software portion on existing processor resources. Leveraging approximately six orders of magnitude performance advantage over pre-silicon simulation, Dacota's post-silicon approach is able to offer significantly higher coverage compared to pre-silicon techniques. The average performance overhead of Dacota, compared to execution with error detection disabled, is 26% for SPLASH2 benchmarks, which exhibit memory access patterns fairly representative of typical programs. Moreover, through reuse of existing hardware resources, Dacota is capable of limiting its area penalty to than 0.01% for a commercial design, such as the OpenSPARC T1 system, and it can provide a two to three orders of magnitude area advantage over other popular post-silicon and runtime solutions for memory consistency.

We found that Dacota is effective in detecting subtle consistency and coherence bugs, showing its promise as a solution to the problem of validating the order of memory operations in CMP systems. Furthermore, Dacota enables post-silicon debugging support, providing invaluable information to the validation team. Once the post-silicon validation phase is completed, Dacota can be disabled before product shipment, thus completely eliminating the experience of any performance degradation by an end-user.

Solutions such as Dacota and Reversi, overviewed in Chapter 3, open new frontiers in microprocessor verification, by allowing functional tests to be executed directly on prototype hardware with significant increase in performance compared to pre-silicon techniques. To become successful, post-silicon techniques must incur minimal hardware overhead and strive to reuse existing on-chip resources for validation purposes, yet still provide sufficient test quality and coverage. Furthermore, promising post-silicon validation approaches should include support for debugging by providing valuable clues to the root cause of a bug. All of these features allow post-silicon validation to shorten the processor verification cycle and eliminate significantly more bugs before the product is released. However, even with these advanced methods, designers cannot completely guarantee correctness of all processor's behaviors, and subtle errors may still slip into production silicon. To combat these escaped errors in hardware components deployed in the field, researchers have began proposing runtime validation solutions, techniques that operate concurrently with the system monitoring the correctness of its functionality and providing correction mechanisms when errors manifest in the field. In the remaining chapters of this book we overview several of the most prominent techniques in this domain, analyzing their performance and bug finding capabilities.

References

[BA08] David Burger and Todd Austin. The SimpleScalar toolset, version 3.0, 2008. `http://simplescalar.com`.

[CMP08] Kaiyu Chen, Sharad Malik, and Priyadarsan Patra. Runtime validation of memory ordering using constraint graph checking. In *HPCA, Proceedings of the International Symposium on High-Performance Computer Architecture*, pages 415–426, February 2008.

[DWB09] Andrew DeOrio, Ilya Wagner, and Valeria Bertacco. Dacota: Post-silicon validation of the memory subsystem in multi-core designs. In *HPCA, Proceedings of the International Symposium on High-Performance Computer Architecture*, pages 405–416, February 2009.

[Knu97] Donald E. Knuth. *The Art of Computer Programming*. Addison-Wesley Professional, third edition, 1997.

[LTS$^+$07] Ana S. Leon, Kenway W. Tam, Jinuk L. Shin, David Weisner, and Francis Schumacher. A power-efficient high-throughput 32-thread SPARC processor. *IEEE Journal of Solid-State Circuits*, 42(1):7–16, January 2007.

[MCE$^+$02] Peter S. Magnusson, Magnus Christensson, Jesper Eskilson, Daniel Forsgren, Gustav Hallberg, Johan Hogberg, Frederik Larsson, Andreas Moestedt, and Bengt Werner. Simics: A full system simulation platform. *IEEE Computer*, 35(2):50–58, 2002.

[MS06] Albert Meixner and Daniel J. Sorin. Dynamic verification of memory consistency in cache-coherent multithreaded computer architectures. In *DSN, Proceedings of the International Conference on Dependable Systems and Networks*, pages 73–82, June 2006.

[MSB$^+$05] Milo M. K. Martin, Daniel J. Sorin, Bradford M. Beckmann, Michael R. Marty, Min Xu, Alaa R. Alameldeen, Kevin E. Moore, Mark D. Hill, and David A. Wood. Multifacet's general execution-driven multiprocessor simulator (GEMS) toolset. *SIGARCH Computer Architecture News*, 33(4):92–99, 2005.

[SLL02] Jeremy Siek, Lie-Quan Lee, and Andrew Lumsdaine. *The Boost graph library: user guide and reference manual*. Addison-Wesley, 2002. `http://www.boost.org/doc/libs/release/libs/graph`.

[SMHW02] Daniel J. Sorin, Milo M.K. Martin, Mark D. Hill, and David A. Wood. SafetyNet: improving the availability of shared memory multiprocessors with global checkpoint/recovery. In *ISCA, Proceedings of the International Symposium on Computer Architecture*, pages 123–134, 2002.

[WOT+95] Steven Woo, Moriyoshi Ohara, Evan Torrie, Jaswinder Singh, and Anoop Gupta. The SPLASH-2 programs: characterization and methodological considerations. In *ISCA, Proceedings of the International Symposium on Computer Architecture*, pages 24–36, June 1995.

Part III
RUNTIME VERIFICATION FOR MODERN MICROPROCESSORS

As microprocessor complexity grows with each new generation of products, the capabilities of even the most advanced pre-release (*i.e.*, pre- and post-silicon) verification solutions still severely lags behind, allowing bugs to slip into production hardware and compromise the stability and security of customers' systems. Because of this, researchers in academia and industry recently started to propose solutions that extend verification and protection from bugs to a deployed processor by means of patching and/or additional on-chip hardware. The purpose of this additional hardware components is to monitor the stability of the system at runtime. In this third part of the book we overview and classify escaped bugs in commercial processors available today, collecting them from publicly available errata report documents. We then present research solutions leveraging hardware checkers to ensure correctness of runtime operation. This introduction is followed by the presentation of a comprehensive solution for patching processors' hardware in the field and for protecting users from unknown escaped bugs. Following the discussion of techniques to protect individual processor cores, we turn our attention to multi-cores and present one checker-based and one patching-based runtime verification solution for these architectures. In our concluding remarks we recap the state-of-the-art approaches to microprocessor validation and outline future research thrusts that are crucial in closing the verification gap and protecting both manufacturers and end-users from the severe impacts of processor design bugs.

Chapter 6
RUNTIME VERIFICATION WITH PATCHING AND HARDWARE CHECKERS

Abstract. With this chapter we begin our discussion of the functional correctness solutions that can be pursued past the release of a new microprocessor design, when the device is already shipped and installed in an end-customer's system. Error detection at this stage of the processor's life cycle entails monitoring its behavior dynamically, by observing the internal state of the device with dedicated hardware components residing on the silicon die. In addition to error detection, runtime validation solutions, also called in-the-field solutions, must include an effective recovery and error bypass algorithm, to ensure minimal performance loss and forward progress of the system even in presence of bugs. To make the case for dynamic validation in this chapter, we first discuss the type, criticality and number of escaped design errors reported in several processors products. We then overview two major classes of runtime solutions: checker-based approaches and patching-based ones: the main difference between these two techniques lies in the underlying error detection mechanism. Checker-based solutions focus on verifying high-level system invariants, usually specified at design-time and then mapped to dedicated hardware components. Patching techniques address bugs of which the manufacturer becomes aware after product release and provide programmable means to describe these bugs so that the system can later identify their occurrence at runtime. We then contrast these two frameworks in terms of error coverage, usage flow and performance overhead, and present in detail some of the most popular academic and industrial solutions known today for each of the two classes.

6.1 Analysis of Escaped Errors in Commercial Processors

Despite impressive efforts by microprocessor manufacturers to build correct designs, bugs do escape the verification process. This section examines escaped errors reported by the respective design house for a number of AMD, Intel (x86 and ARM families), and IBM (PowerPC family) processors. We study errata reports for several designs and classify each escaped bug based on its nature and

I. Wagner and V. Bertacco, *Post-Silicon and Runtime Verification for Modern Processors*, DOI 10.1007/978-1-4419-8034-2_6, © Springer Science+Business Media, LLC 2011

hardware unit affected. The results of our analysis are plotted in Figure 6.1. The data sources for the chart are from several Pentium and Celeron products [DDJ06, Kon09, Int02a, Int02b, Int04, Int05], Athlon and Opteron processors [AMD05], StrongARM-SA1100 [Int00], and PowerPC 750GX [IBM05]. The first observation that can be made from the chart is that the control portion of the design is the source of a fairly large fraction of escaped bugs. Below we discuss each of the categories in our classification:

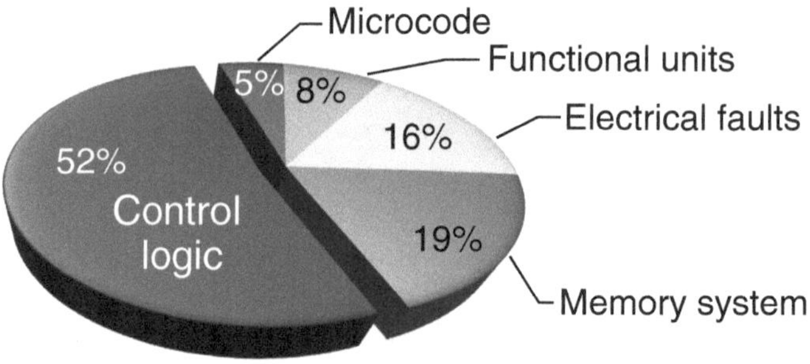

Fig. 6.1 Classification of escaped bugs reported for a number of commercial processors. The chart groups escaped bugs into five different classes and plots the relative incidence of each group. Data sources for the analysis are from a number of x86 [DDJ06, Kon09, Int02a, Int02b, Int04, Int05, AMD05], StrongARM-SA1100 [Int00], and PowerPC 750GX [IBM05] processors.

Processor's control logic: These bugs are the result of incorrect decisions made at the occurrence of important execution events and bad interactions between concurrent events. Usually they are caused by an unforeseen execution scenario and/or incomplete or incorrect functional specification. An example of this type of escape was reported for the Opteron processor, where a "REP MOVS" instruction was reported to cause the following instruction to be occasionally bypassed [AMD05].

Functional units: These are design bugs in functional units that may lead to erroneous computation results. This category includes bugs in microarchitectural components, such as branch predictors and translation look-aside buffers. A well known example of this type of bugs was reported for the Pentium processor, where a lookup table used to implement the division algorithm contained some incorrect entries. This bug, also known as the "FDIV bug", led the processor to compute incorrect results when performing a division between specific pairs of input values [DDJ06].

Memory system control: These are bugs in the implementation of the processor's memory subsystem, which includes caches, memory interfaces and instruction prefetching units. An example of this type of bug was reported for the Pentium III processor, where in certain specific situations the interaction between the instruction fetch unit and the data cache unit through the main bus could cause the entire system to hang [Int05].

Microcode: These are bugs in the implementation of the microcode for particular instructions. These bugs may be as well considered software bugs, since microcode programs for individual instructions are implemented in software and uploaded to dedicated memory units. As a result, once they are known, they may also be corrected more easily than those in the previous categories, through what is called a microcode update. An example of such bugs can be found in the 386 processor, where the microcode responsible for checking the minimum size of the task state segment, which must be 103 bytes long, incorrectly allowed segments of 101 and 102 bytes due to a flaw [Kon09].

Electrical faults: These are design errors occurring when certain logic paths do not meet the specified timing requirement. Usually these paths are only activated under exceptional and rare conditions, which had not been foreseen at design time, and thus had not been checked. In addition, they can only be manifest if a processor is operating at or close to its maximum frequency-voltage point, so that propagation delays are stressing the timing requirements. An example of this problem occurred in the StrongARM SA-1100 processor, where the load register signed byte (LDRSB) instruction would fail when it was fetched from the pre-fetch buffer by not meeting the required time constraint [Int00].

As it can be noted from the chart, control logic bugs dominate the errata reports with a 52% incidence, at least in the systems that we studied. The high occurrence of this type of escape bugs can be justified by considering the design complexity entailed in developing these control logic blocks, which must handle the interactions between multiple instructions. This complexity is responsible for making it impossible to fully verify these modules with formal tools at design time. While these units are thoroughly tested in simulation, frequently only a fraction of their wide range of functional behaviors can be validated due to the short product development timelines and the relatively slow simulation performance for RTL descriptions. As a result, unforeseen corner case situation may remain unchecked. Other related studies on the sources of errors corroborate our findings, including the work by Van Campenhout [CMH00]. This study reports that many design flaws are the result of incorrectly implemented interactions between major processor components or an unexpected combination of rarely occurring events.

A recent analysis of reported escaped bugs by Constantinides, *et al.* [CMA08] indicates that the number of discovered escaped bugs is increasing with each new generation of microprocessors. Assuming that the fraction of bugs discovered after product release is proportional to the number of bugs that are indeed present in the product and to the number of parts deployed in the field, this leads to the conjecture that the number of bugs escaping into finished products is actually increasing. The conjecture can be justified by the increased complexity, particularly in the control logic, of newer processor designs.

Finally, the advent of multi-core architectures seems to negatively affect the overall landscape. Indeed the same study by Constantinides, *et al.* notes that the rate at which new escaped bugs are discovered is three times larger for multi-core systems than it was for even the most complex uniprocessor designs. An independent analysis by DeOrio, *et al.* [DBB08], focusing on multi-core processors errata documents, reveals that 10% of the bugs in these architectures occur in the memory subsystem. Thus, it seems that the most recent processor architectures are manifesting a worsening of overall design quality, possibly due to new, complex memory subsystems. The non-deterministic propagation delay in most memory subsystems of modern multi-cores leads to non-unique, and sometime unpredictable, execution outcomes. These are much harder to verify than the outcome of uniprocessor systems, for which there is only one correct result that can be easily corroborated by an architectural simulator.

All of these studies clearly indicate the inability of pre-release verification solution to produce a processor that is completely bug-free, and such inability is becoming sharper, as each new product generation is released with more and more escaped bugs. It comes with no surprise then that many design houses are becoming more and more interested in investigating runtime, or in-the-field, solutions for functional correctness to address this pressing issue. As of today a few partial solutions have been developed in the industry and are already deployed in available microprocessors. More radical and encompassing techniques have been proposed in academic settings and research labs and, in some cases, are being explored for potential commercial deployment.

6.2 Classification of Runtime Verification Solutions

The runtime verification solutions available in industry or proposed by academia can be grouped into two families: checker-based techniques, as the ones described in [Aus00, MBS08, MS06, DBB07] and patching solutions such as those in [WBA08, SNC$^+$07, GC95, MP99]. The common trait of checker-based techniques is that error detection is performed through a dedicated hardware component; this component is responsible for monitoring the execution continuously at runtime to detect anomalies or violations of invariants and triggering recovery when necessary.

DIVA [Aus00] is an example of one of the first checker-based solutions to ensure correctness of execution for a single core processor. This framework augments a uniprocessor's complex pipeline with a very simple "checker core"; the checker core validates all computation results of the complex core by recomputing the same operations. If the simple core detects a mismatch, its results are assumed correct and fed back to the complex core to recover from the error. A similar solution by DeOrio, *et al.* [DBB07] strives to reduce the area overhead imposed by the DIVA system by limiting the focus of the checker to only detect and correct the control logic errors in the complex pipeline. In this solution, it is assumed that the datapath of the core can be verified formally, since it consists only of the functional units, and excludes all complex inter-instruction activity that is typically coordinated by the control units along with many performance-enhancing units, such as branch predictors, instruction pre-fetchers, *etc.* As a result, the formally verified datapath can be relied upon for error recovery and correction, eliminating the need to dedicate additional hardware for the correct computation of results.

A recent work by Meixner, *et al.* [MBS08], called Argus, proposes another checker-based runtime verification solution for simple processor cores. This mechanism does not focus on functional bugs that escaped design-time verification, but instead targets the protection from soft errors, electrical errors of short duration that may occur when the silicon chip is hit by radiation. Soft errors are becoming an increasing concern in the semiconductor industry because smaller, nanometer scale, silicon transistors are more susceptible to them than previous technology generations. Argus leverages compile time information about a program's execution flow that is embedded in the program's binary code. During program execution, this information is compared against signatures generated at runtime by checkers distributed throughout the core. When an error is detected through a signature mismatch, Argus flushes the pipeline and re-executes the offending instruction.

Patching-based approaches, on the other hand, focus on providing programmable mechanisms to allow a system to side step the occurrence of bugs at runtime. A requirement of patching-based solutions is that bugs must be known by vendors and encoded in the programmable framework in order to be avoided. Thus, processor manufacturers are required to diagnose an escaped bug that they want to correct to determine the system's configurations that trigger it, and create and distribute patches to sidestep those situations among all end-users. Two main techniques in this domain have been developed in industrial settings and are deployed in a number of products. Both use software mechanisms to address functional errors:

Instruction patching. Instruction patching allows to bypass functional errors that affect the implementation of individual instructions. In this technique, the assembly source code of a software application must be inspected, and each occurrence of the corrupted instruction must be replaced with other instructions, or short assembly code sequences, to perform the same operation. By creating software programs that do not use the corrupted instruction, the vendors prevent the functional bug hidden in its implementation from being triggered. Note that this technique requires recompilation of every software application running on the system. Instruction patching was the initial work-around used to bypass the Pentium FDIV bug: Linux- and Windows-based compilers were updated to generate code which would run a preliminary test to determine if the underlying processor suffered from the error. If the test indicated so, a floating-point divide emulation routine would be used instead of FDIV instruction to avoid using the hardware divider [SE04]. A similar technique was also used to port Windows NT to Alpha processors [BHS99]: a hardware bug in the underflow exception mechanism of Alpha processors forced software developers to make the operating system step in and emulate the offending instructions in software. A specific advantage of this approach was that it could operate in a completely transparent fashion to the user – beside the requirement of installing an operating system patch. Performance-wise, however, this approach does not seem very promising. For example, the FDIV fix in the Microsoft Visual C++ compiler may incur up to a 100% performance overhead on a flawed processor. Even on correctly working designs, the workaround may impose up to a 10% penalty [Mic09].

Microcode patching. As the term suggests, microcode patching consists of mechanisms to modify microcode instruction sequences in a system deployed in the field. In this solution, microcode programs are stored in dedicated memory components and can be updated by specialized software patches. Reportedly, both Intel and AMD processors include this ability to update their microcode after deployment in the field [GC95, MP99, Car90]. Microcode patches are loaded into a small on-die buffer memory during system startup. This buffer memory is consulted first when microcode routines are called in the pipeline's decode stage, thus overriding existing microcode. This technique can change the semantics of any individual instruction, similarly to instruction patching; however, the main advantage of this solution is that microcode substitution occurs internally in the pipeline, thus no modification of software applications is necessary. The concept of patchable microcode is not new, as many early computers, such as the Xerox Alto and DEC LSI-11, supported writable micro-stores, allowing engineers to update the implementation of individual instructions at any time [BP80]. Related to microcode patching is also a firmware update technique, which modifies BIOS and motherboard settings with the purpose of correcting functional issues at the system-level. For example, adjusting the processor's voltage and frequency, or disabling certain performance features in the BIOS may prevent the occurrence of certain bugs. One important aspect in development of microcode patching solutions is their security: indeed it is extremely important that only the processor's vendor has the ability to modify the design's microcode; if it was possible for a third party to accomplish the same goal, then systems would be extremely vulnerable to security attack directly at the hardware level.

While these techniques have proven their value in commercial solutions, their potentially high performance impact limits their viability in many situations. For instance, in the case of the Pentium FDIV bug, all divide instructions had to be tested to detect if they manifested the bug, and replaced by emulated routine if needed, leading to significant slowdowns. Additionally, many control logic bugs cannot be easily associated with a particular instruction, thus they cannot be avoided through any of these solutions. For example, the Intel 486 processor included a bug that manifested when a non-maskable interrupt occurred in the same cycle as a global segment violation; in this situation the violation would not be detected [DDJ06]. Short of emulating every instruction, this bug could not be fixed with instruction patches. Motivated by these limitations, researchers have started to propose solutions that allow the processor's control logic units to be patched directly, thus addressing non instruction-centric issues, such as interactions between operations or between instructions and interrupts. These techniques are still very new and are still at the research stage.

In the remainder of this chapter we discuss and compare a number of runtime verification solutions: those by Austin [Aus00], DeOrio, *et al.* [DBB07], Meixner, *et al.* [MBS08] and Sarangi, *et al.* [SNC$^+$07]. We start by investigating DIVA – one of the first dynamic verification approaches based on a complete functional checker. We also examine the technique proposed by DeOrio, *et al.* [DBB07], which leverages design-time formal verification to reduce the checker area, and compare it against the DIVA solution. Next, we present Argus [MBS08], a solution for checker-based runtime verification of simple processor cores. The work of Sarangi, *et al.* [SNC$^+$07] is a patching-based approach proposing a distributed patching mechanism and several error recovery and bypass schemes.

In the upcoming Chapter 7 we will continue our discussion of hardware patching solutions with the presentation of two additional techniques recently developed by us: Field-Repairable Control Logic (FRCL) [WBA08] and Semantic Guardians [WB07]. Finally, in Chapter 8 we shift our focus to runtime verification of multi-core systems, in particular, to in-the-field validation of such communication invariants as cache coherence and memory consistency. There we discuss a checker-based solution developed by Meixner, *et al.* [MS06] and our own patching approach, called Caspar [WB09]. Both are recent proposals to combat a growing number of memory subsystem bugs and ensure correctness of execution in the context of multi-cores.

6.3 DIVA: Dynamic Verification of Microprocessors

The DIVA project [Aus00] by Austin constitutes one of the first runtime verification solutions for microprocessor designs. Considered one of the key contributions in the field of dynamic validation, the project introduced the term *dynamic hardware verification* and sparked interest for this field in many research environments, both in industry and academia. The work makes a strong case for validation beyond traditional pre-release methods, because of the inability of design-time verification tools to keep up with the growing complexity of processor architectures. It also argues that any practical runtime technique must have very small area and performance impacts in order to be viable in commercial solutions, which are very sensitive to cost and user experience.

DIVA (Dynamic Implementation Verification Architecture) consists of a hardware component, called the "checker core", implementing a simple in-order processor pipeline, and deployed next to a complex uniprocessor core, whose functionality was not exhaustively validated at design time. Figure 6.2 shows a conceptual schematic of a DIVA-augmented processor. The task of the checker core is to validate the correctness of each of main core's computations at runtime, by detecting and correcting erroneous execution and allowing only error-free operations to retire.

When the checker core detects an error in the execution, it raises an exception and replaces the incorrect value with an internally computed correct result. Instructions that were in flight in the main core are then flushed, so that any effects of the erroneous computation are eliminated. The main core then restarts operation from the last known-correct architectural state. The situation is somewhat similar to a misprediction in a speculative branch operation: the main core in DIVA "predicts" that all operations will complete without errors, while the checker verifies the prediction before allowing an instruction to retire, and flushes the pipeline if a "misprediction" occurs. To ensure forward progress in the pipeline and to detect deadlock errors, the checker core also includes a watchdog timer. The timer is set to the maximum theoretical latency of any operation, and it is reset whenever an instruction is committed.

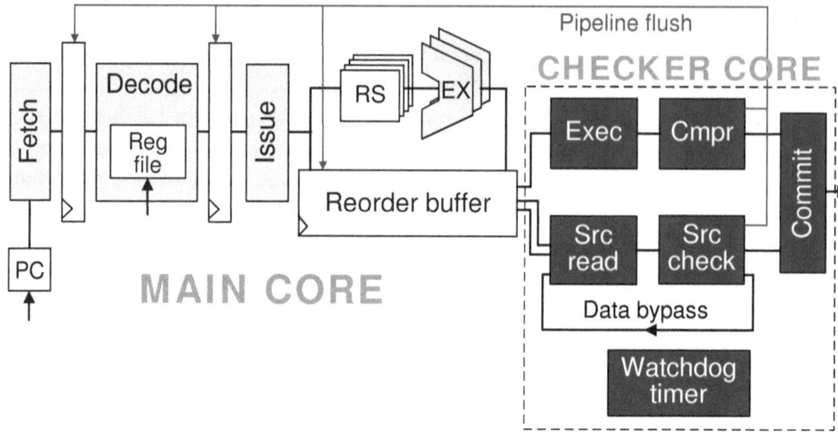

Fig. 6.2 DIVA architecture. DIVA augments a complex uniprocessor pipeline (main core) with a simple in-order processor (checker core). The checker verifies the correctness of each dynamic instruction, allowing only error-free operations to retire. For each instruction, it checks that the correct input values were used and recomputes the instruction's results. Note how error-free computation and correct inter-instruction communication are validated in concurrent stages of the checker pipeline. If a mismatch is detected, the main core's computation is discarded and the checker's results are committed. The watchdog timer ensures forward progress and detects pipeline deadlocks.

6.3.1 Checker core operation

The checker core enforces several crucial invariants of program execution in modern processors, that is, *correctness of computation, control* and *communication*, as well as guarantee of *forward progress*. The watchdog timer enforces forward progress, while the checker core enforces the other invariants by recomputing instructions' results and verifying correctness of communication between operations.

For each instruction being executed, the main core supplies the checker with source values and the result that it had computed. With this data, the checker re-executes the instruction internally (stage *Exec* in Figure 6.2) and the outcome is compared to the one provided by the main core (stage *Cmpr*). This activity of the checker core allows to validate results of arithmetic operations, as well as target address computations in control instructions. Note that the computation units in the checker may be implemented differently than the main core's arithmetic and logic units. For instance, the ALU in the checker should be designed to minimize area and maximize verifiability, since DIVA must rely on the functional correctness of this unit. On the other hand, the ALUs in the main core should prioritize performance.

Communication-related errors may arise due to memory address miscalculations, renaming logic faults, *etc.* To provide comprehensive protection against such errors, the DIVA checker has access to the main core's architected registers and memory (stage *Src read*) and verifies that values used by the main core as instruction inputs are indeed correct (stage *Src check*). These two checker stages operate concurrently with the stages validating the computation, so that the overall checker pipeline is only two stages deep. Frequently, instructions executing in the main core may receive their input values via the bypass/forwarding network from preceding instructions. As it can be gathered from Figure 6.2, however, by the time an instruction reaches the *Src read* stage, all preceding operations – with the exception of the one in the *Src check* stage – have completed and committed their results to the proper registers and/or memory locations. In the *Src read* stage, the checker core reads the values from the specified input locations, which by now have been updated and compares them with values used by the main core. Consequently, DIVA can validate that any data forwarding and reordering that has occurred in the main core adheres to the architectural communication model. For instructions depending directly on values computed in the preceding instruction, DIVA relies on a simple bypass bus from *Src check* to *Src read*. Finally, DIVA assumes that individual storage elements in the core are protected by error-correcting codes (ECC), so that no errors may occur while storing new values. If a communication error is discovered, the main core is flushed and the instruction is re-executed: during the second execution the main pipeline should retrieve the source values from architected storage, thus the new result should be free of communication-related errors.

One of the critical aspects in the DIVA solution is the difference in performance between the main core and the checker core. The former is optimized for performance and it includes such features as register renaming, out-of-order execution, branch prediction, *etc.* The checker core is designed for verifiability and minimal area impact, thus it values simplicity over performance. However, the checker core is capable of trailing the main core's operation at the same speed because it can benefit from all the results computed by the main core, and it simply validates them. It would be possible to view the DIVA solution as comprising a simple processor (the checker core) connected to an extremely complex and highly accurate speculative unit (the main core), which provides the correct information (results, destination addresses and registers, *etc.*) most of the time. The DIVA checker effectively "rides in the wake" of the main pipeline's execution, observing the speculatively computed values and control flow transitions produced by it. This creates a large amount of instruction-level parallelism in the checker, drastically simplifying the complexity of its circuitry. Consequently, the checker is amenable to formal verification approaches, which guarantee absence of functional bugs in that part of the design and, by extension, correctness of operation of the entire device.

6.3.2 DIVA in action

We now illustrate the operation of a processor design equipped with a DIVA checker through an example. Consider a system as the one shown in Figure 6.3.a: there are five instructions currently in flight in the main core; three additional instructions have completed their execution and are now in the checker core for validation.

For sake of the example, assume that instruction *INST2* executed incorrectly. The checker's *Cmpr* stage detects the incorrect result and raises the pipeline flush signal

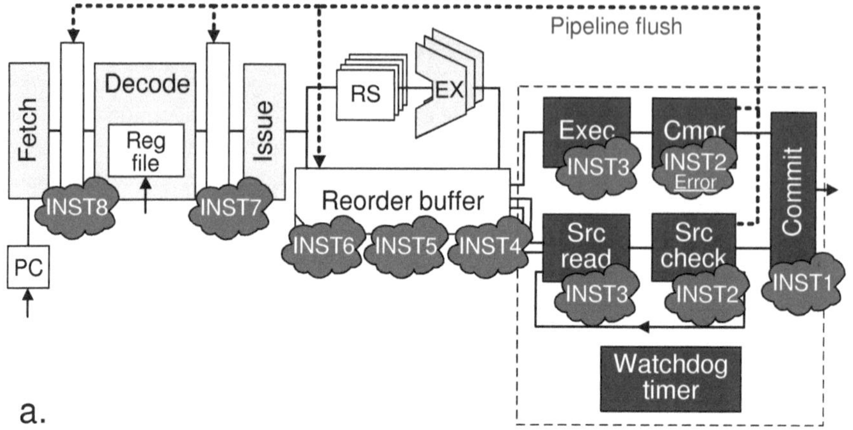

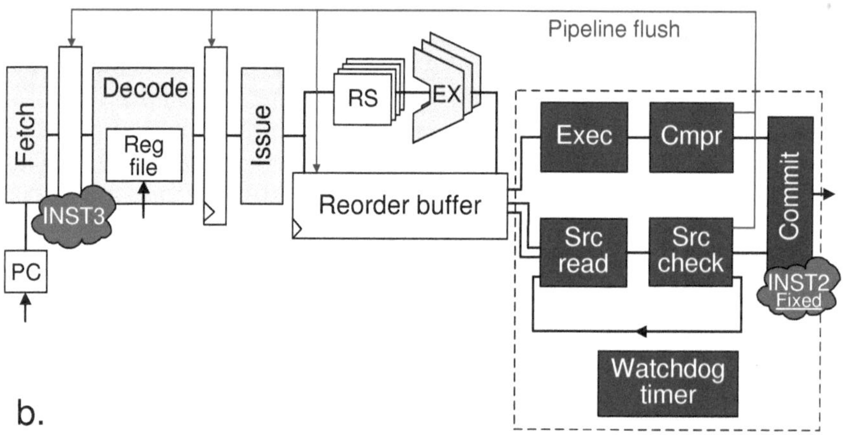

Fig. 6.3 Example of DIVA operation. A complex uniprocessor core equipped with a DIVA checker core is executing an instruction stream. **a.** Three instructions have completed execution and are undergoing validation in the checker core before being committed. After the results for *INST2* are recomputed in the checker's *Exec* stage, the *Cmpr* stage detects an error by comparison with the original results, and triggers a pipeline flush to initiate recovery. **b.** *INST2*'s result has been corrected and the instruction is committed. The pipeline is then restarted at *INST3*, so to guarantee that the incorrect result generated by *INST2* does not propagate to subsequent operations.

to initiate a recovery. At the time of the detection, instruction *INST1* was already in the commit stage, thus it has updated the architectural storage and it has completed. Instruction *INST2* will be corrected in the *Cmpr* stage and the correct result will be used in the commit stage. Instruction *INST3* and subsequent ones will restart execution after the pipeline flush is completed, as shown in Figure 6.3.b. The re-execution of all instructions following instruction *INST2* is necessary so to prevent the original invalid result from propagating to subsequent computations.

6.3.3 Benefits and limitations

In addition to functional errors, the DIVA mechanism is also very effective against transient faults (soft errors) that may occur in the main core. As mentioned in Section 6.2 during the discussion of the Argus solution, latest generations of nanometer scale transistors are increasingly susceptible to these issues, due to the smaller channel size. The incidence of radiation causing soft errors is uniform over the chip's surface, thus the large main core is much more prone to this type of errors than the small checker core. As a result, DIVA's ability to protect the main core from errors of this nature, together with the protection provided by ECC-enhanced storage elements, provides a fairly effective solution against computation errors due to transient faults.

Note that DIVA can also be deployed for post-silicon validation support. In this setting, the checker is used to detect errors while the system is running complex testbenches in pre-release hardware prototypes. More interestingly, the checker core is extremely valuable in providing a workaround for critical bugs detected in post-silicon, allowing the validation process to continue without requiring a costly and time-consuming design respin to fix the issue.

From the author's evaluation in [Aus00], the performance impact of the checker core on the overall system is minimal. In absence of errors, the throughput of the system is mostly unchanged, since the checker core simply increases the system's latency. There is, however, an estimated 3% cycles-per-instruction (CPI) penalty due to the access to registers and memory by the checker core in its *Src read* stage. These accesses compete for the same resources as the main core's accesses to register file and memory. If additional read ports were available for the checker core's accesses the penalty would become negligible (0.03%). When errors are present, the performance impact is directly related to the errors' frequency of occurrence: since each error triggers a pipeline flush, there is a performance cost in refilling the system's pipeline. For instance, with errors occurring every 1000 cycles, the performance impact was measured to be approximately 2.6%, while if recovery had to be performed for each instruction, the device could only operate at 7% of its original speed while still producing functionally correct results.

The area impact of the checker core, on the other hand, can be an important limitation of DIVA framework. This core must feature a functionally complete processor and it is estimated to occupy approximately 10% of the total die area. This issue has been addressed by DeOrio, *et al.* in [DBB07], who observed that that errors in functional units have become relatively rare, while control logic bugs are prevalent. This can be attributed to the fact that historically formal verification tools have been used predominantly to verify functional units and datapath modules with well defined behavior. In contrast, control logic blocks change radically from generation to generation of processors, and their complexity continues to increase as designs implement always new and advanced technologies, such as multi-threading and virtualization. Focusing on the higher benefit of protecting control logic, DeOrio, *et al.* have suggested a runtime verification scheme with significantly lower area footprint than DIVA. The smaller area footprint is a direct benefit of a smaller checker block, which trusts the correctness of the main core's datapath.

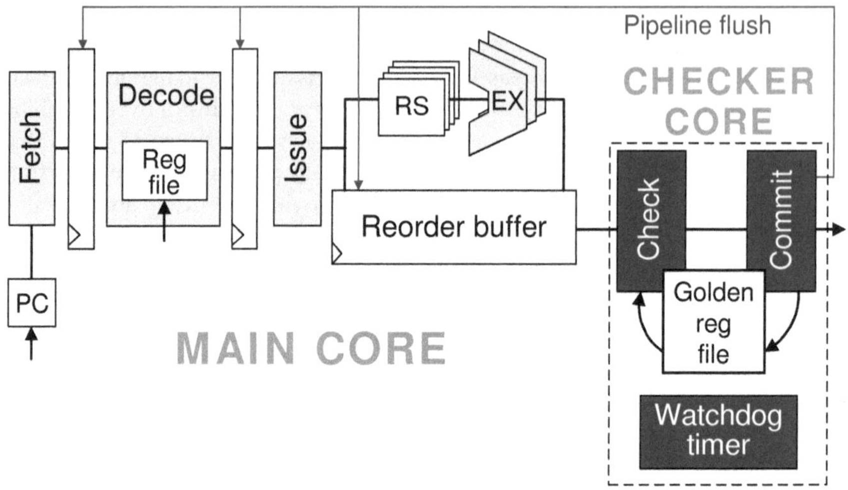

Fig. 6.4 Chico architecture for control logic bugs. A complex processor pipeline is augmented with a checker block comprising a *Check* and a *Commit* stages, a golden register file and a watchdog timer. Upon instruction commit, values are simultaneously updated on both the main core's and the golden register file. The *Check* stage verifies that register renaming and data forwarding in the out-of-order core were performed correctly. When an error is detected, the pipeline is flushed and the main core is reconfigured to a simple mode of operation, allowing only one instruction in the pipeline at a time. Correctness of execution during error recovery is guaranteed because input values are always retrieved from the register file. Normal operation may resume once the offending instruction completes.

The authors of [DBB07] assume that the processor's datapath has been formally verified at design time, thus the computation results by the processor core are always correct. Errors may still occur because of potential bugs hidden in the design's control logic, which may cause incorrect input values to be sent to the execution units. The solution proposed by DeOrio, *et al.*, called Chico, augments the main core with a checker block comprising a *Check* and a *Commit* stages responsible for validating the correctness of value propagation to and from execution units. In addition, the checker core includes a dedicated golden register file and a watchdog timer as shown in Figure 6.4. The golden register file is used to validate inter-instruction communication in the main core, and the watchdog timer guarantees forward progress as in the DIVA solution.

When instructions are committed, values are updated in both register files simultaneously. Note that the main core's register file is often designed specifically for out-of-order execution and register renaming, thus contains several times more registers than are visible to the software. The golden register file, on the other hand, only contains the architected registers, and, consequently, it is much smaller and simpler. Instructions are propagated through the pipeline together with their source values, received either from the register file or from the forwarding network. In the *Check* stage, these values are compared to those stored in the golden register file to verify the correctness of renaming logic, data forwarding, *etc.* An error in the renaming logic or in the bypass network would manifest as a mismatch in the *Check* stage of the Chico core, triggering the recovery process. When recovery is needed, the processor's pipeline is flushed and reconfigured to a simple mode where only one instruction is allowed in the main core at a time. As a result, in this mode of operation all control logic errors are circumvented, since input values are simply retrieved from the register file and the formally verified datapath ensures that computation executes correctly. After the failed instruction is re-executed, the processor may resume normal operation until the next mismatch occurs or, as in DIVA, a watchdog timer detects a pipeline deadlock.

Although conceptually similar to DIVA, the approach by DeOrio, *et al.* is different in several key points. First, it has lower coverage, since only data communication errors can be detected by the checker. Additionally, while the DIVA checker contains the recovery mechanism in itself, Chico relies on existing processor units that are amenable to formal verification, namely the datapath. By trading off a fraction of the coverage while employing formal verification on a portion of the main core, the authors were able to significantly reduce the area of the checker, making it an order of magnitude smaller than the full fledged multi-stage checker pipeline in DIVA.

6.4 Runtime Verification of Simple Cores with Argus

As mentioned in Chapter 4, with the advent of multi-core architectures over the past ten years, there has been a shift from extremely complex and deeply pipelined uniprocessors chips to simpler and smaller cores to be deployed in several units on a multi-core silicon die. In the 90's, wide superscalar, out-of-order pipelines were the staple of processor design; by the end of the decade, their excessive power density ultimately had put an end to the "megahertz race" and had forced manufacturers to investigate more energy efficient architectures. Today, the prevalent solution is a processor with several relatively simple and efficient processing elements, communicating with each other via on-die interconnect. Multi-core designs have become mainstream of the industry for a wide range of market segments: from supercomputers to handhelds and smartphones. Individual cores in these processors usually have single-issue, in-order pipelines, approximately five to fifteen stages deep. The requirements for effective dynamic verification of these architectures are then fairly different than those of the complex out-of-order cores targeted by DIVA. Indeed, the DIVA checker was in itself a simple pipeline re-executing the instructions completed by a complex uniprocessor. In a multi-core environment, augmenting the individual cores with a DIVA-style checker would be inefficient in terms of die area, power and performance.

A runtime verification solution optimized specifically for simple processor cores was presented very recently in the work by Meixner, *et al.* [MBS08]. The framework, called Argus, combines several dynamic verification techniques to provide a comprehensive solution validating dataflow, control flow, computational correctness and memory accesses in simple cores. A signature-based dynamic flow technique proposed by the authors in their earlier work [MS07] is used to check data flow and control flow. Computational correctness is checked by a set of distributed functional checkers, which augment individual blocks of the execution engine in the processor. Finally, correctness of memory operations is verified with a novel address/data hashing scheme, capable of detecting misdirected accesses, as well as data corruption. Note that the Argus solution focuses on error detection alone, and it does not provide mechanisms for runtime error correction. Hence, to bypass detected errors, must rely on instruction re-execution or emulation. As a result, Argus can provide near-full protection against transient faults caused by SEUs and voltage fluctuations, but it has limited potential to recover from functional design errors.

The basis of Argus' error detection mechanism is a technique that establishes equivalence between static data and control flow graphs and the dynamic flow of the program being executed. To compute static flow graphs, Argus operates at compile time, dividing the binary program into basic blocks, usually at granularity of branch instructions, and generating signatures representing the flow within each block. The dataflow graph consists of a set of source and sink nodes representing the machine's architectural state at the beginning and end of an instruction block, intermediate nodes representing individual instructions and directed edges representing data/control dependencies. The signature of an instruction within a flow is derived by hashing the instruction's ID and its source registers; it is then appended to the instruction's destination register. Through these registers signatures are propagated and encapsulated to all instructions in the basic block. Final signatures for architectural state registers are concatenated together to form the block's dataflow graph signature (DGS). The DGS's are embedded into the binary program either with special instructions or by reclaiming unused bits in native instructions and will be used at runtime for dynamic verification.

To conduct runtime verification, statically computed DGS's are extracted from the program's binary and are compared to signatures dynamically generated at runtime. To this end, each architectural register in the processor's pipeline is augmented with dedicated storage for signatures. Signature are propagated together with register values through the pipeline, where they are updated dynamically by the execution units and memory interface modules, and then are written back to the dedicated storage, as illustrated in Figure 6.5. Program counter and memory have also signature registers associated with them. The former allow to detect errors in control flow transfer, *i.e.,* when the program counter is incorrectly modified by a branch instruction; the memory signature is handled slightly differently and ensures that all operations are delivered correctly to the memory system. Note that Argus does not track dataflow through individual memory locations, since this would require prohibitively large amounts of overhead for signature storage. At the end of the execution of each basic block, the Argus checker compares the extracted compile-time

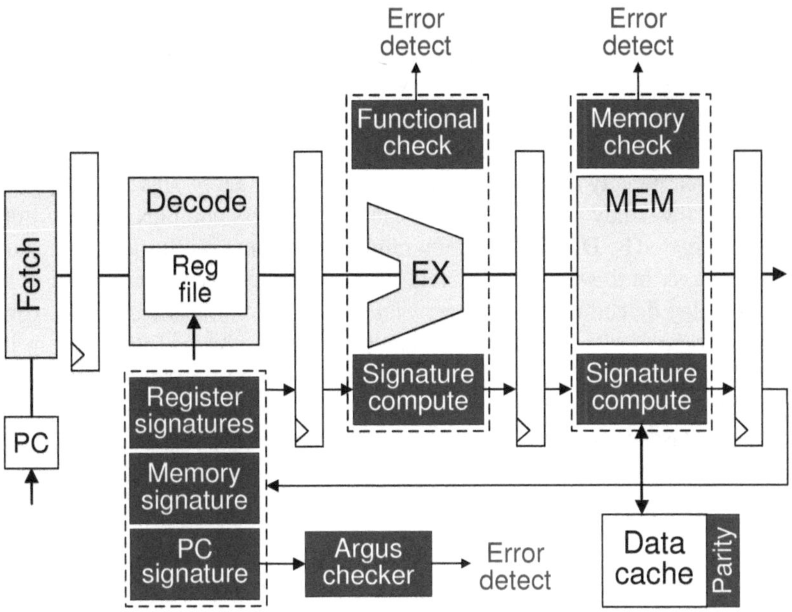

Fig. 6.5 Argus architecture. In Argus, architectural state elements (register file, PC, memory) are augmented with signature registers. The values of these registers are computed dynamically for each executed instruction and associated with the instruction's destination register. The Argus checker compares these signatures with the ones computed statically and embedded in the program's binary to detect possible dataflow errors. Additionally, correctness of computation is verified by functional checkers shadowing main units, while memory errors are detected by parity error detection codes.

signature with the one calculated dynamically, to verify the correctness of the execution flow. In addition to the current block's DGS, Argus piggybacks the signatures of the block's legal successors to the program binary. Therefore, transitions between blocks due to data dependent branches and jumps can also be verified dynamically.

In addition, Argus checks the actual computation of individual instructions with distributed checkers, shadowing the processor's main functional units. These checkers are specifically designed to perform simple and fast validation of the main blocks by re-computing only a subset of bits of the result. The number of bits computed by the checkers can be selected by the designer, allowing to trade off error coverage for checker's area and performance. Issues such as data corruption and misdirected accesses are monitored through parity codes and a dedicated address/data multiplexing technique. In this technique, for each store access, the physical address of the location is *xor*-ed with the data and written to memory along with the parity bit of the data. On a read, the physical address computed by the load instruction is *xor*-ed with the loaded value and parity is checked, to detect bit flips and other memory subsystem errors. Finally, to provide the ability to detect potential deadlocks, Argus augments the design with a watchdog timer guaranteeing that the pipeline is not stalled for longer than the programmed number of cycles.

Experimental results demonstrate that Argus detects as much as 98% of unmasked transient and permanent faults injected into the gate-level implementation of an OpenRISC 1200 processor core. A fraction of the masked faults, *i.e.,* errors that do not compromise the correctness of execution, was also detected by Argus; more importantly, the framework never produced a false positive. The authors estimate the area penalty of the approach to be approximately 10%. Recall that the main core in this study was significantly smaller and less complex than the out-of-order core targeted by DIVA, thus maintaining a small area profile was a much more challenging task in this project. Overall, Argus provides excellent protection from transient faults, detection of permanent faults and a flexible tradeoff of error detection coverage vs. design area/performance. While the Argus framework in itself does not provide a recovery mechanism beside program re-execution, the solution could be augmented with formally verified rollback and recovery hardware, such as the solutions discussed in Chapter 7, to provide a general and efficient solution for functional runtime verification.

6.5 Hardware Patching Approaches for Runtime Verification

As mentioned in Section 6.2, hardware patching solutions present an alternative to checker-based techniques in ensuring correctness of processor operation at runtime. Indeed, patching approaches rely on a pool of on-die programmable monitors, which continuously observe the internal state of the design, in contrast with checkers solutions employing re-computation and verification of global invariants. In a patching framework, a hardware error must first be detected and reported to the manufacturer. Here, in turn, the support team investigates the issue, diagnoses the triggering conditions of the bug and encodes it into a patch. The patch is then distributed to end-users and uploaded to the on-die programmable monitors at system startup. At runtime, when an internal configuration of the processor matches the scenario described in the patch, a recovery mechanism is initiated to bypass a potential occurrence of the error. Compared to checker-based solutions, patching techniques usually incur significantly smaller area and performance overheads in absence of bugs; on the other hand, they provide protection only for known and debugged issues. As discussed earlier, patching has been traditionally applied at the instruction or micro-operation level in practical industry solutions (see instruction and microcode patching in Section 6.2). Yet, these industry-deployed approaches are not all-powerful: they miss a large portion of subtle processor bugs, particularly ones related to interaction between instructions at runtime. Moreover, patching may occasionally may cause severe performance loss. To address the growing need for more efficient and rigorous solutions, researchers have proposed patching techniques that operate on a level lower than microcode, and, therefore, allow for a more precise identification of errors. One such technique was developed concurrently by a group of researchers at the University of Illinois – Urbana-Champaign and at the University of California – San Diego [SNC+07].

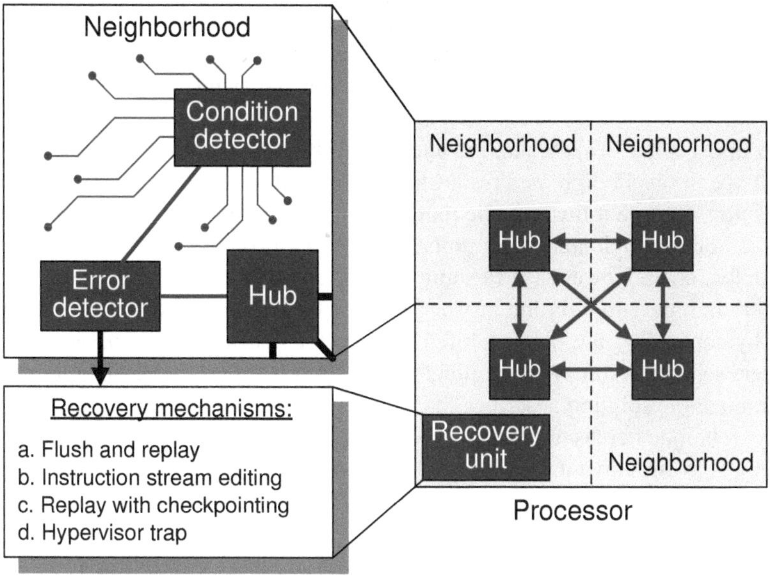

Fig. 6.6 Architecture for runtime hardware patching The processor die is partitioned into "neighborhoods", each containing a condition detector and an error detector unit. Condition detectors tap into wires within a neighborhood and monitor them at runtime for specific, pre-programmed conditions. Error detector blocks are responsible for identifying if an observed sequence of events leads to an erroneous state of the system and recovery is required. Erroneous states spanning multiple neighborhoods can be detected with the support of a programmable communication network, linking all detectors on the processor die together. Possible recovery mechanisms include pipeline flushing and re-execution (with or without checkpointing), instruction stream editing and hypervisor-based recovery.

In this solution, the authors recommend the use of a field-programmable on-die error detection network, consisting of multiple *condition detectors* tapping into signals within a region (called a "neighborhood") of a processor die. A neighborhood in this case may include the processor core and other on-die peripheral controllers. During runtime execution, these detectors observe the values of the tapped signals and match them against pre-programmed conditions. In an event that a condition, such as a specific state transition or signal value change, is encountered, the condition detector alerts the corresponding error detector, which is pre-loaded with *error fingerprints*. In the framework of [SNC$^+$07], each fingerprint consists of a set of events and time intervals between them to describe a hardware bug. Thus, at runtime, error detectors keep track of encountered conditions and their time stamps, and compare them to error fingerprints in the patch. Note that several error detectors, each observing a small block in the design, can operate in parallel, ensuring simultaneous identification and avoidance of multiple hardware bugs. Moreover, the authors discuss the possibility of interconnecting detectors in different regions of the processor via a dedicated network to allow for efficient diagnosis of the most complex processor errors (see Figure 6.6). Once all conditions and time intervals of a bug are matched, detectors alert the processor recovery unit, which can be configured to invoke one or several of the circumvention mechanisms discussed below.

Flush and replay. This technique attempts to bypass hardware errors by squashing all operations in flight and restarting execution from the first not-yet-committed instruction. Note that flushing the pipeline alleviates pressure on the execution engine, retirement logic and other units of the CPU: this by itself may be sufficient to avoid an error. However, if flushing alone is incapable of bypassing the bug, the recovery unit may modify the instruction issue timing or the interleaving of operations by controlling the pipeline's fetch stage (see *instruction stream editing*. This recovery scheme is somewhat similar to the degraded mode of operation of our field-repairable logic solution described in Chapter 7. However, unlike that approach, the replay technique here is not formally verified to provide reliable error bypass, rather it attempts to circumvent the error probabilistically. If the issue is not resolved during replay, a more sophisticated recovery method is invoked at the next detection.

Instruction stream editing attempts to bypass errors by dynamically altering the stream of operations executed by the pipeline when a bug is detected. To this end, *noop* instructions can be inserted into the fetched stream to modify the interleaving between conflicting operations. Note that the operation stream is edited only when a conflict leading to an error is detected. Therefore, this technique avoids the unnecessary overheads of microcode patching, which replaces each occurrence of an instruction with an altered micro-operation stream, regardless of conflicts.

Replay with checkpointing is an extension of the *flush and replay* method that can be useful in situations where the architectural state of the machine is corrupted by an erroneous activity. For example, an improperly issued store instruction may modify the state of the caches in a multi-core machine; this must be properly restored before recovery through flush and replay commences. To enable this solution, the authors propose to augment the design with a checkpointing framework such as SafetyNet [SMHW02] or ReViveI/O [NMGT06]. We will overview checkpointing techniques in Chapter 8 in the context of runtime validation of multi-core processors. Note that, in order to make this recovery option available in a system, the checkpointing mechanism has to actively running at all times. This may incur additional costs in silicon area and performance, even when no error is present.

Hypervisor-based recovery is a more sophisticated error bypass scheme, compared to the ones discussed so far, because it builds on a recent trend towards virtualization technology. Hypervisors [PG74, BDF$^+$03, McG07], or virtual machine monitors, are low-level software systems that can execute directly on the computer hardware; they provide an API so that guest operating systems and applications may access the hardware. In other words, hypervisors take the place of the main operating system and serve as "drivers" for the processor, memory and peripherals that programs use to execute on the machine. Note, however, that virtual machine monitors do not generally emulate execution, but instead serve applications' requests for hardware management in a broad sense. In a system supporting virtualization technology, when an error detector matches an error fingerprint, a hypervisor can be invoked to bypass the issue. Depending on the designer choice, the hypervisor may provide the required service through controlled rollback, hardware-reconfiguration, emulation of a portion of the program, or other situation-specific solutions.

After investigating bugs reported in errata of various commercial processors, Sarangi, *et al.* claim that the majority of those issues can be corrected at runtime with their patching solution, with a negligible overhead due to wire-tapping and fingerprint storage. The authors also suggest that performance deterioration due to increased wire load and the need to conduct multi-cycle fingerprint matching is minor, and can be recovered with additional design-time circuit optimizations. Another important step, that must be taken during the design process to enable this solution, is the selection of wires to be monitored in the processor, a crucial aspect for precise and efficient error identification. This selection, however, must be completed before any errors in the product are identified. As a result designers are forced to reason about the importance of various internal signals and weigh the benefits and drawbacks of monitoring them. To support this process, the authors suggest the use of an algorithm that can automatically select signals based on past error history and processor layout [STT06].

An alternative approach for providing runtime functional error detection through patching was proposed very recently by Constantinides, *et al.* [CMA08]. Instead of selecting specific signals to monitor, as in [SNC$^+$07], this work provides a novel framework where *every* storage element in the design is monitored. To this end, all standard flip-flops in the system are extended with additional latches to store condition patterns and with combinational logic to perform the matching. The matching flip-flops are divided into segments, based on their physical location on the processor die. All those belonging to a same segment are then tied together with *or* gates to provide a segment-wide matcher. These segment-wide matchers are then connected through a tree-like network to the central unit, which is responsible for initiating recovery and bypass when an error is detected. Note that in this case every element of the processor's state is monitored, thus, the coverage of the design's internal state in this framework is greatly increased and engineers do not need to pre-select which signals to observe at design time. For comparison, in the work of Sarangi, *et al.* 270 of the Pentium 4's internal signals are observed, while in [CMA08] 39,000 flip-flops in OpenSPARC T1 were monitored. However, the area overhead of the solutions grows correspondingly, surpassing 15% of the OpenSPARC T1 processor die. Furthermore, the sizes of individual error signatures and patches to be distributed to customers are also larger. Nevertheless, the approach developed by Constantinides still holds much potential for systems where reliability and correctness are paramount, *e.g.,* aero-space and military applications. It may also become viable for consumer microprocessors-based systems in the future, as transistors shrink and silicon real estate becomes less expensive.

In the next chapter we overview in detail another patching solution that predates [SNC+07]. In developing that approach, called field-repairable control logic (FRCL), we relied on the observation of the internal processor state and on a comparison with pre-loaded bug patterns. Unlike the technique of Constantinides, we deem the area overhead of the solution to be a crucial factor, and thus limit state monitoring to a few crucial control logic signals to minimize silicon cost. The matcher in our technique is monolithic, as opposed to the distributed and networked error detector units in both frameworks described in this section. In our experimental evaluation we found that, even in fast, out-of-order pipelines, the propagation delay through the undivided matcher does not exceed the critical path delay of the main core, allowing us to forgo a tree- or network-based interconnect. The matcher in our work is optimized specifically for the detection of pipeline control errors: the dominant type of processor bug as shown in Section 6.1. In addition, we describe a novel recovery mechanism: a formally verified, reliable mode of pipeline operation that allows us to safely bypass control logic bugs and ensure forward progress. Finally, we analyze an extension of FRCL for hardware-security assurance applications, called a *semantic guardian*. Unlike patching solutions, which can only protect from known and diagnosed errors, the semantic guardian framework ensures correctness of operation even against hidden, unknown bugs. Combined with FRCL programmable matchers, semantic guardians enable a flexible and versatile protection from a variety of design errors in modern complex microprocessors.

6.6 Conclusions

Despite the best effort of pre-silicon verification and post-silicon validation, bugs still slip into commercial hardware, leading to a wide range of potential problems, from security holes in the end-user systems to safety issues, and to substantial financial impacts on vendor companies. As indicated by the analysis in this chapter, the majority of these escaped bugs occur in the control logic blocks of modern microprocessors and can be attributed to the extreme complexity of these modules. To cope with the growing problem of escaped bugs, vendors traditionally relied on system-level patching solutions, which modify the software, BIOS or microcode executed by the processor to bypass hardware errors. Unfortunately, these techniques often incur a significant performance overhead and cannot be used to fix all control logic bugs. Therefore, researchers have recently proposed alternative solutions for runtime verification, based on hardware checkers and patching. One of the most prominent of the former group is DIVA, which augments a complex processor pipeline with a simpler processor core to re-execute all operations and dynamically verify the correctness of both control logic and datapath. In this chapter we also overviewed two other checker-based solutions, one targeting the correctness of just the control logic portion of a processor (DeOrio, *et al.*) and the other focusing on the runtime validation of simple processor cores (Argus). We noted, however, that all hardware-checker based solutions incur a fairly high area overhead due to the necessary inclusion of one or more checker modules. In contrast, hardware patching solutions also discussed in this chapter have lower area penalty and, therefore, are more likely to be adopted by the industry in the near future. In the next chapter we discuss hardware patching with the field-repairable control logic (FRCL), analyze its performance and area impact o this approach, and suggest how this technique can be extended to protect even against unknown escaped errors.

References

[AMD05] Advanced Micro Devices, Inc. *Revision Guide for AMD Athlon™64 and AMD Opteron™Processors*, August 2005. `http://www.amd.com/us-en/assets/content_type/white_papers_and_tech_docs/25759.pdf`.

[Aus00] Todd Austin. DIVA: A dynamic approach to microprocessor verification. *Journal of Instruction-Level Parallelism*, 2:1–26, May 2000.

[BDF+03] Paul Barham, Boris Dragovic, Keir Fraser, Steven Hand, Tim Harris, Alex Ho, Rolf Neugebauer, Ian Pratt, and Andrew Warfield. Xen and the art of virtualization. In *SOSP, Proceedings of the ACM Symposium on Operating Systems Principles*, pages 164–177, October 2003.

[BHS99] Eric B. Brett, David P. Hunter, and Sharon L. Smith. Moving Atom to Windows NT for Alpha. *Compaq DIGITAL Technical Journal*, 10(2):1–26, January 1999.

[BP80] Henry Baker and Clinton Parker. High level language programs run ten times faster in microstore. *ACM SIGMICRO Newsletter*, 11(3-4):171–177, 1980.

[Car90] Adrian Carbine. *U.S. Patent no. 5253255: Scan mechanism for monitoring the state of internal signals of a VLSI microprocessor chip.* Intel Corporation, November 1990.

[CMA08] Kypros Constantinides, Onur Mutlu, and Todd Austin. Online design bug detection: RTL analysis, flexible mechanisms, and evaluation. In *MICRO, Proceedings of the International Symposium on Microarchitecture*, pages 282–293, November 2008.

[CMH00] David Van Campenhout, Trevor Mudge, and John P. Hayes. Collection and analysis of microprocessor design errors. *IEEE Design and Test of Computers*, 17(4):51–60, 2000.

[DBB07] Andrew DeOrio, Adam Bauserman, and Valeria Bertacco. Chico: An on-chip hardware checker for pipeline control logic. In *MTV, Proceedings of the International Workshop on Microprocessor Test and Verification*, pages 91–97, December 2007.

[DBB08] Andrew DeOrio, Adam Bauserman, and Valeria Bertacco. Post-silicon verification for cache coherence. In *ICCD, Proceedings of the International Conference on Computer Design*, pages 348–355, October 2008.

[DDJ06] DDJ Microprocessor Center, 2006. `http://www.x86.org/`.

[GC95] Michael D. Goddard and David S. Christie. *U.S. Patent no. 5796974: Microcode patching apparatus and method.* Advanced Micro Devices, Inc., November 1995.

[IBM05] International Business Machines Corporation. *IBM PowerPC 750GX and 750GL RISC Microprocessor Errata Notice*, July 2005. `http://www-01.ibm.com/chips/techlib/techlib.nsf/techdocs/CC0918EAD88CB09E87256F5C006FF9E6`.

[Int00] Intel Corporation. *Intel®StrongARM®SA-1100 Microprocessor Specification Update*, February 2000.

[Int02a] Intel Corporation. *Intel®Celeron®Processor Specification Update*, September 2002. `http://developer.intel.com/design/celeron/specupdt/24374847.pdf`.

[Int02b] Intel Corporation. *Intel®Pentium®II Processor Invalid Instruction Erratum Overview*, July 2002. http://developer.intel.com/design/pentiumii/specupdt/24333749.pdf.

[Int04] Intel Corporation. *Intel®Pentium®Processor Invalid Instruction Erratum Overview*, July 2004. http://www.intel.com/support/processors/pentium/sb/cs-013151.htm.

[Int05] Intel Corporation. *Intel®Pentium®III Processor Specification Update*, May 2005. http://download.intel.com/design/PentiumIII/specupdt/24445355.pdf.

[Kon09] Dusko Koncaliev. Bugs in the Intel microprocessors, 2009. http://www.cs.earlham.edu/~dusko/cs63/.

[MBS08] Albert Meixner, Michael E. Bauer, and Daniel J. Sorin. Argus: Low-cost, comprehensive error detection in simple cores. *IEEE Micro*, 28(1):52–59, 2008.

[McG07] Harlan McGhan. The gHost in the machine. *Microprocessor Report*, 21(3):11–33, March 2007.

[Mic09] Microsoft Corporation. */QIfdiv (Enable Pentium®FDIV Fix)*, 2009. http://msdn2.microsoft.com/en-us/library/ms856573.aspx.

[MP99] Kevin J. McGrath and James K. Pickett. *U.S. Patent no. 6438664: Microcode patch device and method for patching microcode using match registers and patch routines*. Advanced Micro Devices, Inc., October 1999.

[MS06] Albert Meixner and Daniel J. Sorin. Dynamic verification of memory consistency in cache-coherent multithreaded computer architectures. In *DSN, Proceedings of the International Conference on Dependable Systems and Networks*, pages 73–82, June 2006.

[MS07] Albert Meixner and Daniel J. Sorin. Error detection using dynamic dataflow verification. In *IPACT, Proceedings of the International Conference on Parallel Architectures and Compilation Techniques*, pages 104–118, September 2007.

[NMGT06] Jun Nakano, Pablo Montesinos, Kourosh Gharachorloo, and Josep Torrellas. ReViveI/O: efficient handling of I/O in highly-available rollback-recovery servers. In *International Symposium on High-Performance Computer Architecture*, pages 203–214, February 2006.

[PG74] Gerald J. Popek and Robert P. Goldberg. Formal requirements for virtualizable third generation architectures. *Communications of the ACM*, 17(7):412–421, 1974.

[SE04] Amitabh Srivastava and Alan Eustace. ATOM: A system for building customized program analysis tools. *ACM SIGPLAN Notices*, 39(4):528–539, April 2004.

[SMHW02] Daniel J. Sorin, Milo M.K. Martin, Mark D. Hill, and David A. Wood. SafetyNet: improving the availability of shared memory multiprocessors with global checkpoint/recovery. In *ISCA, Proceedings of the International Symposium on Computer Architecture*, pages 123–134, 2002.

[SNC$^+$07] Smruti Sarangi, Satish Narayanasamy, Bruce Carneal, Abhishek Tiwari, Brad Calder, and Josep Torrellas. Patching processor design errors with programmable hardware. *IEEE Micro*, 27(1):12–25, 2007.

[STT06] Smruti Sarangi, Abhishek Tiwari, and Josep Torrellas. Phoenix: Detecting and recovering from permanent processor design bugs with programmable hardware. In *MICRO, Proceedings of the International Symposium on Microarchitecture*, pages 26–37, December 2006.

[WB07] Ilya Wagner and Valeria Bertacco. Engineering trust with semantic guardians. In *DATE, Proceedings of Design, Automation and Test in Europe Conference*, pages 743–748, April 2007.

[WB09] Ilya Wagner and Valeria Bertacco. Caspar: Hardware patching for multi-core processors. In *DATE, Proceedings of Design, Automation and Test in Europe Conference*, pages 658–663, April 2009.

[WBA08] Ilya Wagner, Valeria Bertacco, and Todd Austin. Using field-repairable control logic to correct design errors in microprocessors. *IEEE Transactions on Computer-Aided Design of Integrated Circuits and Systems*, 27(2):380–393, 2008.

Chapter 7
HARDWARE PATCHING WITH FIELD-REPAIRABLE CONTROL LOGIC

Abstract. This chapter describes in detail field-repairable control logic (FRCL), a solution that we recently developed [WBA06]. FRCL is a patching-based runtime verification technique that relies on an on-die programmable matcher. The matcher compares the state of the processor to patterns describing known bugs. In particular, FRCL targets control bugs in microprocessor cores, which, as our analysis in Section 6.1 shows, dominate the landscape of escaped errors in commercial products. To detect such control bugs, the matcher is engineered to monitor multiple critical signals in a core's control logic block. The patterns stored in the matcher are developed by the manufacturer after an escaped error is detected and diagnosed; patterns are then distributed to end-users via patches, such as BIOS updates. When a buggy situation is detected, the matcher recovers from it with a pipeline flush and invokes a degraded mode of operation. In this mode, the complexity of the processor is greatly reduced, sacrificing some performance features, but allowing to formally prove the complete functional correctness of the system. Once the buggy situation is bypassed, the processor resumes normal high-performance operation. We analyze different aspects of FRCL operation and describe a methodology for automatic selection of signals to be monitored by the matcher. Finally, we extend the field-repairable control logic framework with semantic guardians - hardware circuits encoding all control states of the design that have been verified prior to its release. With the help of the guardians the processor can be guaranteed to always operate in a verified state (in either normal or degraded mode), thus enabling trusted computation.

7.1 Introduction

Once a system is verified to a satisfactory level during its development, a design house will ramp up its production and start distributing it to customers. Unfortunately, due to shrinking development timelines and rapidly increasing complexity of modern processors, released designs are never fully and exhaustively verified and often contain subtle errors. These *escaped bugs* are eventually identified when

I. Wagner and V. Bertacco, *Post-Silicon and Runtime Verification for Modern Processors*,
DOI 10.1007/978-1-4419-8034-2_7, © Springer Science+Business Media, LLC 2011

the system is already deployed in the field and cannot be easily corrected by the manufacturer, since they require modification of the actual silicon die. Recognizing the inevitability of such errors and the need to efficiently fix them without a costly product recall, researchers in recent years started to develop hardware patching solutions. Our analysis in Section 6.1 demonstrates that the majority of such escaped errors occur in the control logic portion of the design, which traditionally has been the hardest to verify. To enable efficient in-the-field patching of a processor's control logic, this chapter presents a low-cost and expressive mechanism, called field-repairable control logic (FRCL), originally presented in [WBA06] and later extended in [WBA08]. In this framework, when an escaped bug is exposed in the field, the support team investigates it and generates a pattern describing the control state of the processor that triggers the occurrence of the bug. The pattern is then sent to end customers as a patch and is loaded into an on-die *state matcher* at system startup. The matcher constantly monitors the state of the processor and compares it to the stored patterns to identify when the pipeline has entered a state associated with a bug. Once the matcher has determined that the processor is in a flawed control state, the pipeline is flushed and forced into a *degraded mode* of operation for the execution of the subsequent instruction.

In degraded mode, the processor starts execution from the first uncommitted instruction and allows only one operation to traverse the pipeline at a time. Therefore, much of the control logic that handles interactions between operations can be turned off. The resulting system is greatly simplified, enabling a complete formal verification of the degraded mode at design time. In other words, we can guarantee that instructions running in this mode complete properly, and thus can ensure forward progress, even in the presence of design errors, by simply forcing the pipeline to run in degraded mode. After the error is bypassed in degraded mode, the processor returns to *high-performance* mode until the matcher finds another flawed control state. In designing the state matcher, we have devoted special attention to creating a system that can detect multiple design errors with minimal false positive triggering. In addition, for situations where the number of patterns of design errors exceeds the capacity of a given matcher, we developed a novel compression algorithm that compacts the erroneous state patterns while minimizing the number of false positives introduced by this process.

In Section 7.4 we extend the ideas of the field-repairable control logic technology further, presenting a solution that provides protection even from unknown escaped bugs in deployed systems. Instead of requiring a vendor company to identify an escaped bug before we can repair it, we simply assume now that *any configuration that was not verified at design time is potentially a buggy configuration*. Thus, in our semantic guardian solution, we protect a system against all those configurations that were not verified at design time. We will show in Section 7.5.6 that we can indeed deliver such protection at minimal performance costs, since, even if the unverified portion of a design is extensive, unverified configurations tend to occur rarely at runtime. Moreover, we discuss how we can match a large set of unverified configuration with little area overhead.

7.2 Field-Repairable Control Logic Overview

This section presents the use flow and the process for correcting escaped bugs for a design incorporating field-repairable control logic (FRCL) technology. We also show the structure of the state matcher circuit and present a pattern compression algorithm for cases when the number of patterns exceeds the size of the matcher. Finally, we analyze an example of an actual bug that is repaired using our approach.

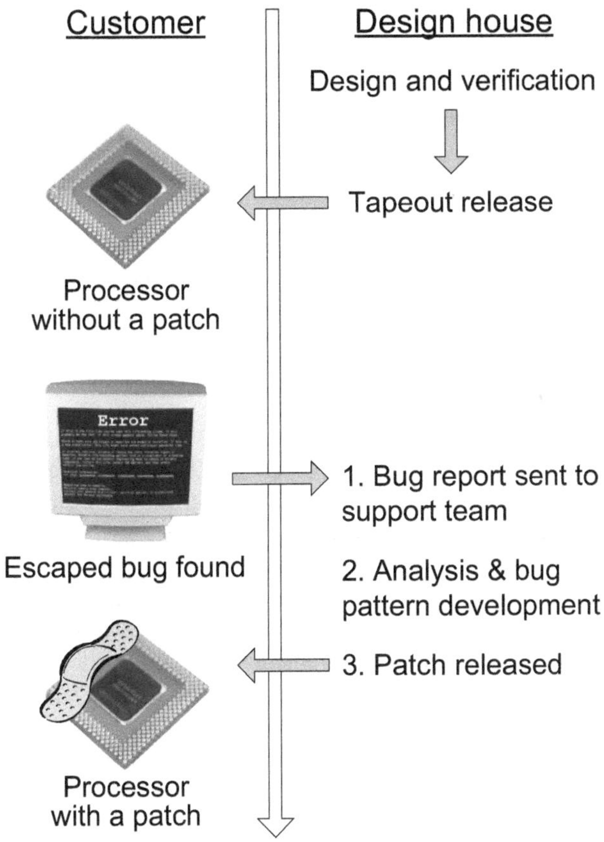

Fig. 7.1 Field-repairable control logic use flow. After a component is shipped to customers and a new bug is exposed in the field, a report detailing the bug is sent to the support team. The error is analyzed and patterns representing the control states associated with the bug are issued as a patch. On every startup, the processor loads the patterns into the state matcher and, if a bug is encountered at runtime, it is bypassed through the reliable degraded mode.

Field-repairable control logic is designed to handle flaws in processor control circuitry for components already deployed in the field. The flow of operation that we envision for this approach is summarized in Figure 7.1 . When an escaped error is detected by end customers, a report containing the error description, such as the sequence of executed operations and the values in the status registers, is sent to the design house. Engineers on the product support team investigate the issue, identify the root cause of the error and which products are affected by it, and decide on a mechanism to correct the bug. As was mentioned in previous chapters, instruction or microcode patching are valid approaches; however, they can have a very high performance overhead or be too costly. We propose that engineers use instead our solution, field-repairable control logic. By knowing the cause of the bug and which signals are monitored by the matcher in defective processors, the engineering team can create patterns that describe the flawed control state configuration leading to the bug. The patterns then can be compressed using the algorithm presented in Section 7.2.3, and sent to customers as a patch. Patches are loaded into the *state matcher*, which is a part of the system, at startup. Every time the patched error is encountered at runtime, a recovery is initiated via degraded mode, as detailed in Section 7.2.4, effectively fixing the bug. From a user standpoint this process is very similar to that of patching software applications: errors discovered in the field are eliminated through software patches sent to the customer base. These patches modify the software application to fix the bug. Likewise, in field-repairable control logic, hardware bugs can be fixed after product release via software patches sent to customers.

7.2.1 Pattern Generation

Patterns to address a design error can be created from the state transition graph (STG) of a device. The correct STG consists of all the legal states of operation, where each state is a specific configuration of internal signals, that are crucial to the proper operation of the device. In addition, these states are connected by all the legal transitions between them. Within this framework an error may occur because of an additional erroneous transition from a legal state to an illegal state that should not be part of the STG, or when an invalid transition connects two legal states, or by the lack of a transition that should exist between states (Figure 7.2). In our solution we add hardware support that relies on *patterns* to detect both illegal states and legal states that are sources of illegal transitions. A pattern is a bit-vector representing the state configuration of the internal signals associated with erroneous behavior of the processor. Note that in this framework a single bug may be represented by multiple patterns if it is caused, for example, by multiple illegal states. To limit the number of patterns required to represent a bug or a set of bugs, we incorporated a novel pattern compression algorithm, presented in Section 7.2.3, into our technology. In a real-world scenario, after receiving a bug report, a product support team would analyze the issue, try to reproduce the error in-house and identify its root cause. Tools such as trace minimizers can be very helpful for this analysis, since they can significantly shorten a trace that leads to a bug, thus simplifying the task of reproducing the problem. Moreover, some of these tools, for example Butramin [CBM05], have the ability to investigate alternative simulation scenarios that reach the same bug. This allows the support team to pinpoint multiple processor control states associated with the bug and identify how these states map to the critical signals monitored by the matcher at runtime. Afterwards, the configurations of the critical control signals leading to the bug are compactly encoded and issued as a patch to end customers. This process is repeated when new design errors or new scenarios exposing known bugs are discovered.

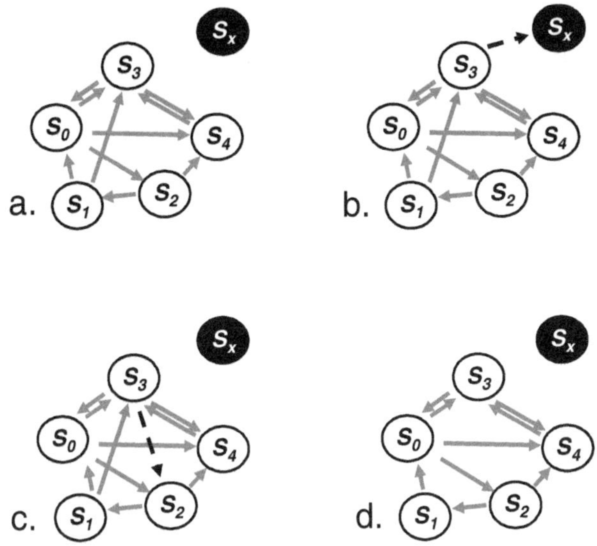

Fig. 7.2 Representing an error through a state transition graph (STG). a. Correct STG of a device. (S_x is an unreachable illegal state). **b.** Erroneous STG including a transition to an illegal state S_x **c.** Erroneous STG including an illegal transition between legal states S_3 and S_2 **d.** Erroneous STG lacking a legal transition, $S_1 \rightarrow S_3$.

7.2.2 Matching Flawed Configurations

As mentioned above, design errors and patterns describing them in our framework are defined through configurations of control signals of the processor and transitions between these configurations. At runtime these signals are continuously observed by a *state matcher* and compared to pre-loaded patterns describing bugs. Therefore, only bugs that manifest via these critical signals can be detected by the matcher. Ideally, all of a design's control signals could be used for this purpose; however, complexity and stringent timing constraints of modern devices prevent such extensive monitoring, allowing only for a small portion of the actual control state to be routed to the matcher. In Section 7.3.3 we present techniques to select these critical state bits among the prohibitively large control state of a microprocessor.

The state matcher can be thought of as a fully-associative cache with the width of the tag being equal to the width of the critical control state vector, which in our

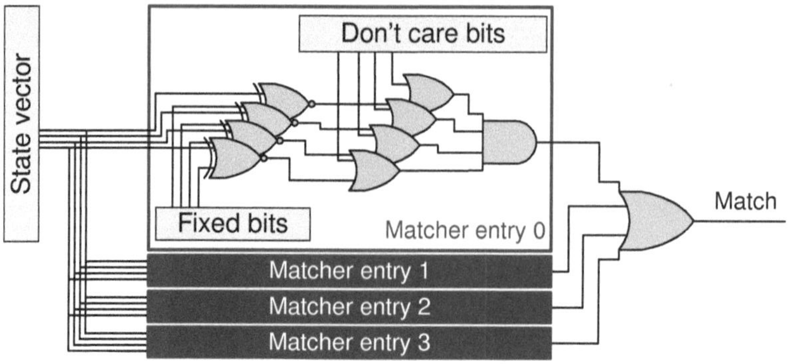

Fig. 7.3 State matcher structure. The critical control state vector is first compared against the fixed bits of a bug pattern. Then the don't care bits in the pattern are overlayed, and the result is reduced to a single match bit. The state matcher comprises multiple independent entries to allow for multiple simultaneous comparisons against many patterns.

experiments was just several tens of bits long. The tag in this case is the pattern describing an erroneous configuration, thus if the control state vector matches any tag in the cache, then a hit occurs and a potential bug situation is recognized. In order to reduce the storage space in the matcher we structured it to allow the use of don't care bits in the patterns to be matched. The don't care bits enable compact representations of multiple individual patterns of the critical control state that differ in just a few bit positions. Using our state matcher, designers issuing a patch can specify a bug pattern through a vector of 0's, 1's and don't care bits (x): 0's and 1's represent the fixed value bits, while x's can match any value in the corresponding control vector position. Note, however, that the control state monitored by the matcher at runtime contains only fixed bit values 0 and 1. Figure 7.7 shows several examples of bug patterns loaded into a four-entry state matcher.

We also anticipate that a single patch may consist of multiple patterns, since a single bug may be associated with several erroneous configurations, as mentioned above, or the design may contain multiple unrelated bugs. To handle this situation the state matcher we developed includes multiple independent entries, as shown in Figure 7.3. On startup, each of the matcher's entries is loaded with an individual pattern containing fixed bits and don't cares. At runtime the matcher simultaneously compares the actual critical control bit values to all of the valid entries and asserts the "Match" signal if at least one entry matches the control state vector. The number of entries in a matcher has to be established at design time and it is one of the engineering design tradeoffs. A larger matcher can be loaded with more patterns, however, it occupies a larger area on the die and has a longer propagation delay. A smaller matcher, on the other hand, might not be able to contain all of the patterns and compression may be needed.

7.2.3 Pattern Compression Algorithm

The pattern compression algorithm that we developed is inspired by classic two-level logic minimization techniques, as described in [McC56]. Our algorithm compresses a number k of patterns into a state matcher with r-entries, where $k > r$. This process, however, often over-approximates the bug pattern and introduces *false positives*, *i.e.*, error-free configurations that will be mis-classified as buggy, and incur some performance impact. Nevertheless, this compression may be necessary to fit several patching patterns into an available state matcher of smaller size.

To map k patterns into an r-entry matcher, the algorithm first builds a *proximity graph*. The graph is a clique with k vertices, once for each of the k patterns, and weighted edges connecting the vertices. The weights on the edges are assigned using a variant of the Hamming distance metric. Specifically, we use an additive metric, whereby corresponding bits are compared one to one, and each $0-1$ pair contributes 1 to the weight, while each $1-x$ or $0-x$ pair contributes 0.5 to the weight. Matching pairs ($0-0$, $1-1$, and $x-x$) do not contribute to the weight. As an example, consider the two patterns $101xx1$ and $1001x1$ shown in left part of Figure 7.4. The two leftmost and two rightmost bits of the patterns are identical, thus they contribute 0 to the weight. Bit 3 of the patterns, on the other hand, form a $0-1$ pair, contributing 1 to the weight, while bit 4 form a $x-1$ pair, leading to a total weight of 1.5 on the edge connecting these patterns. The reasoning behind this weighing structure is fairly straightforward: if we were to compact the two patterns connected by an edge, we would have to replace every discording pair ($0-1, x-0$, and $x-1$), with an x, basically creating a minimum common pattern that contains both the original ones. Matching pairs, however, would retain the values they had in the original patterns. For instance, for the two patterns $101xx1$ and $1001x1$ mentioned above, the common pattern is $10xxx1$, since we have two discording pairs in the third and fourth bit positions. With this algorithm, each $0-1$ pair contributes the same degree of approximation in the resulting entry generated. However, pairs such as $1-x$ or $0-x$, will only have an approximating impact on one of the patterns (the one with the 0 or 1), leaving the other unaffected, hence the corresponding weight is halved.

An exception to the above metric is a case when one pattern is a subset of another pattern. This is possible, because we allow patterns to have don't care bits, essentially representing both 0 and 1 values. In our framework we set the distance between such proximity graph vertices to -1, guaranteeing that these vertex pairs will be chosen first for compression, and the more specific pattern eliminated from the graph.

Once the proximity graph is built, the two patterns connected by the minimum-weight edge are merged together. If $r \leq k$, the compression is completed, otherwise the graph is updated using the compressed pattern just generated, instead of the two original ones, and the process is repeated until we are left with a number of patterns that fits in the matcher.

An example of pattern compression is provided in Figure 7.4. Here, for simplicity, we assume that the matcher can only fit two pattern entries and initially there are

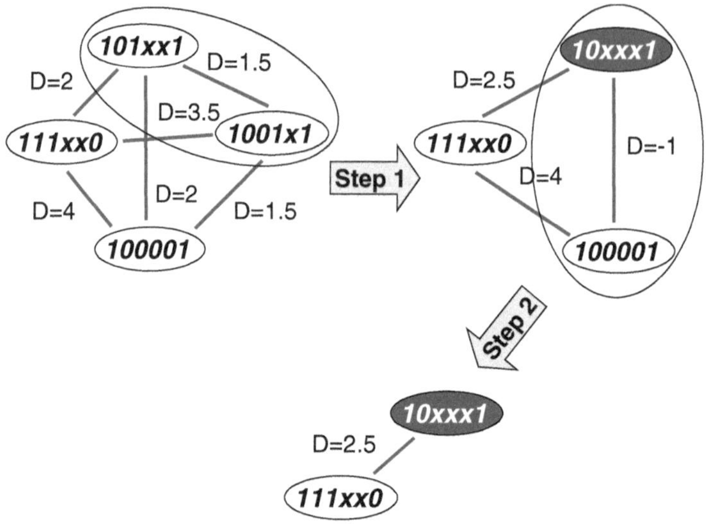

Fig. 7.4 Pattern compression example. Four bug patterns are compressed to fit into a two-entry matcher. A complete graph of the initial four patterns is generated and edges are labeled with our Hamming distance variant metric. The first compression step combines two of the closest (in terms of distance) patterns *101xx1* and *1001x1*. The resultant pattern *10xxx1* has fixed bits in every position where original patterns were identical and don't care bits (x) in all other positions. In the second step, edge weights from the newly generated pattern (shown in a dark gray oval) are recomputed, and then pattern *100001* is eliminated, since it is contained in the pattern *10xxx1*, as the -1 label indicates. The final set of patterns is shown in the bottom part of the figure.

four bug patterns. After the proximity graph is initially built and edges are labeled, the algorithm selects the edge with the smallest distance ($D = 1.5$) and merges patterns $101xx1$ and $1001x1$ connected by it. As discussed above, the resulting pattern is $10xxx1$. When the edge weights are updated after this first step, the graph includes three vertices and it is still too large for the matcher. Note, however, that the newly added pattern ($10xxx1$), contains pattern 100001, thus the edge between them is labeled with distance -1. When the algorithm searches for the edge with the smallest weight for the second compression iteration, this edge is selected and vertex 100001 is eliminated. Compression then terminates, since the resulting set of patterns fit into a two-entry matcher.

Figure 7.5 shows a pseudo-code for the pattern compression algorithm just described. Lines 2 to 8 generate the initial proximity graph by computing the weights of all the edges either by detecting that vertex i contains vertex j (*Contains* function) or computing our Hamming distance variant using the algorithm presented earlier (*Compute_Distance* function). Lines 10 to 13 select the pair to merge, remove one pattern from the set and update the graph. The procedure is repeated until we reach the desired number of patterns. Function *Merge* in line 11, generates a pattern that is the minimum over-approximation of the two input patterns. The function first must check for containment, in which case it returns the container pattern. Otherwise,

```
1   Pattern_Compress()
2   Graph G
3   Foreach Pattern i in G
4    Foreach Pattern j != i in G
5     If Contains(i, j)
6      Weight(i, j) = -1
7     Else
8      Weight(i, j) = Compute_Distance (i, j)

9   While (Num_Patterns > Matcher_Entries)
10   (i, j) = Find_Minimum_Weight_Edge (G)
11   Pattern i = Merge(Pattern i, Pattern j)
12   Delete Pattern j
13   Update_Edge_Weights (G)
14   Num_Patterns--
```

Fig. 7.5 Pattern compression algorithm. Initially, a proximity graph for all bug patterns is generated and edge weights are computed in lines 2-8. The two closest patterns are merged and the graph is updated in lines 10-13. The cycle is repeated until the set of patterns can fit into the fixed size matcher.

it computes the merged pattern by substituting an x bit for each non-matching bit pair. It is worth noting that the performance of the algorithm described could be optimized in several ways, for instance by eliminating all edges with $D = -1$ in the graph at once.

As mentioned earlier, the compression algorithm generates a set of patterns that *over-approximates* the number of erroneous configurations. The resulting set of patterns is still be capable of flagging all the erroneous configurations, however, it will also flag additional correct configurations that have been included by the merging function (*false positives*). The impact on the overall system will not be one of correctness, but one of performance, particularly if the occurrence of the additional critical control configurations is frequent during a typical execution. We measure the amount of approximation in the matcher's detection ability as its *specificity*. The specificity is the probability that a state matcher will not flag a correct control state configuration as erroneous. Specificity can also be thought of as $1 - false\_positive\_rate$. Hence when there is no approximation, the matcher has an ideal specificity of 1; increasing over-approximation produces decreasing specificity values. It is important to note that by virtue of our design and the pattern compression algorithm, our system never produces a false negative, that is, it never fails to identify any of the bug states observable through the selected critical control signals.

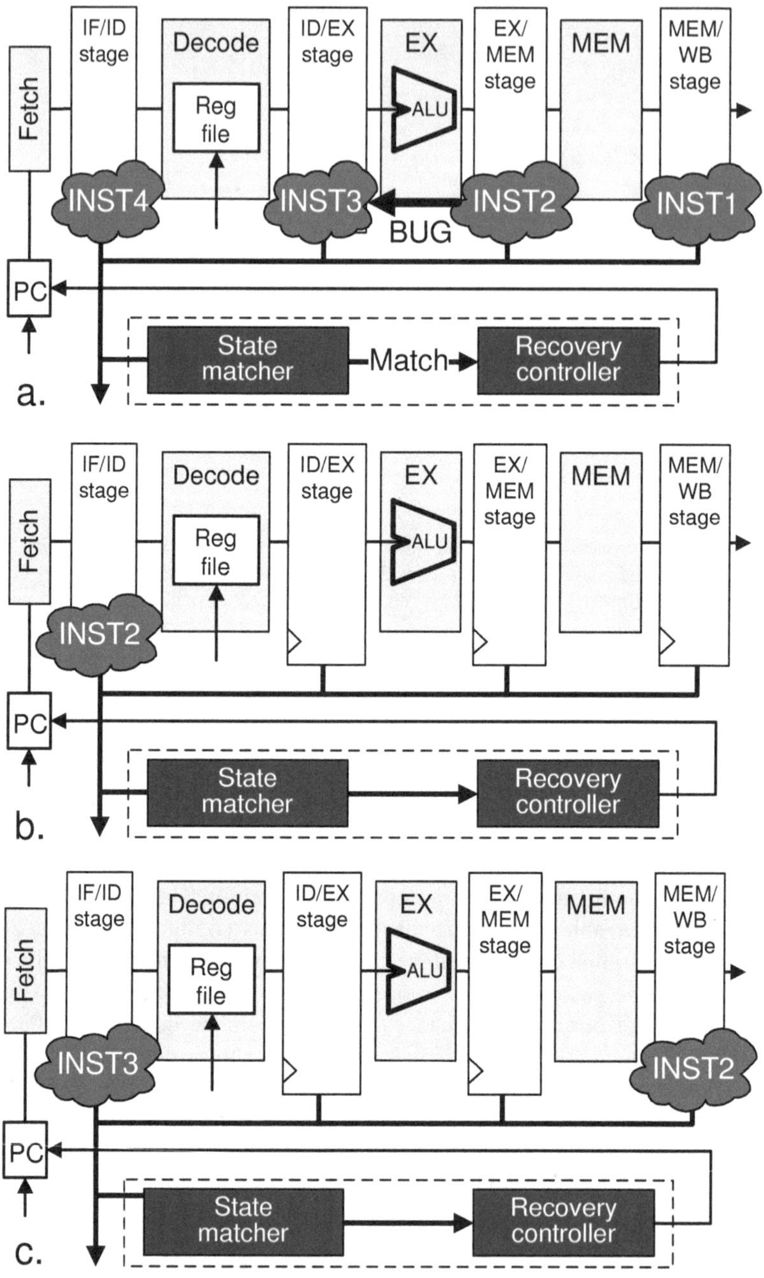

Fig. 7.6 Field-repairable control logic (FRCL) in operation. **a.** The state matcher detects the occurrence at runtime of a control state associated with a bug described in a pre-loaded pattern. **b.** The pipeline is flushed to a known state. Then, execution advances in degraded mode allowing only one instruction in the pipeline at a time. The degraded mode is formally verified, thus guaranteeing forward progress and correctness. **c.** After the offending instructions are bypassed, the high-performance mode of operation is restored.

7.2.4 Processor Recovery

After the set of patterns is generated and possibly compressed, it is issued to end customers as a patch. We envision this step as similar to current microcode patching, where processor patches are included into BIOS (basic input-output system) updates. Updates are distributed by operating system and hardware vendors and are saved in non-volatile memory on the motherboard. At startup, when BIOS firmware executes, patches are loaded into the processor by a special loader. Field-repairable control logic can use an almost identical mechanism, and we expect FRCL patches to be approximately of the same size of a microcode update ($\sim$2KB or less). After the patches, that is the bug patterns, are loaded at startup into the matcher, the processor starts running. While none of the configurations recorded in the matcher occurs during execution, activity proceeds normally (we call this mode of operation *high-performance mode*). However, when a buggy state is detected, the pipeline is flushed and the processor is switched to a reliable degraded mode of execution. Figure 7.6 shows an example of the execution flow occurring a bug pattern is matched in a FRCL-equipped processor. In the example, we consider a simple in-order single-issue pipeline, and we further assume that the interaction between a particular pair of instructions *INST2* and *INST3* triggers a control bug, which has been identified and encoded in a pattern already uploaded in the matcher.

The example shows that, when the pattern is detected by the matcher (Figure 7.6.a), the pipeline is flushed (Figure 7.6.b), that is all instructions except for INST1, which is committed, are squashed. The processor then switches to the degraded mode of execution. This mode is formally verified at design time; hence, we can rely on it to correctly complete the next instruction (INST2), over several execution cycles. Finally, the high-performance mode of operation is restored (Figure 7.6.c) and the pipeline starts to refill starting with INST3. Without FRCL technology, a problem such as the one just described would probably have required re-writing the compiler software or the microcode-related to the instructions, to circumvent the bug configuration. Note that to guarantee correctness with field-repairable control logic it is sufficient to complete only one instruction before re-engaging normal operation since, if the pipeline were to step again into an error state right away, it would once again enter the degraded mode to complete the following instruction. On the other hand, a designer may choose to run in degraded mode for several instructions to guarantee bypassing the bug entirely in a single recovery.

7.2.5 Example

We now show the use of field-repairable control logic through an example inspired by the Intel Celeron bug reported in [DDJ06], which we adapt here, for practical reasons, to a simple five stage pipeline. The processor for the examples includes a design flaw because of which it does not always enforce a required stall cycle between two successive memory accesses. A stall is necessary in these situations

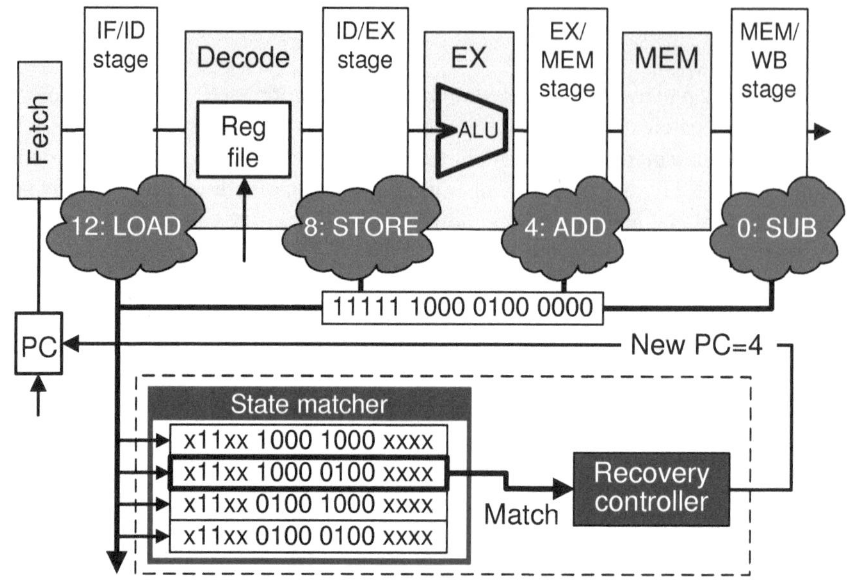

Fig. 7.7 Example of FRCL activity during a memory access bug. The bug in this processor allows two consecutive memory accesses (such as 8:*store* and 12:*load*) to proceed back to back in the pipeline, instead of adding one stall cycle to allow for completion of each memory access. When the bug is recognized by the state matcher, the pipeline is flushed and execution restarts in degraded mode at the first uncommitted instruction (4:*add*). In degraded mode instructions go through the pipeline one at a time: after the load has completed normal mode can be resumed, and, thus, the bug is bypassed.

since all memory operations executes over two cycles: in the first one the memory address is placed on the bus, while on the second the data to/from memory is made available on the same bus. Note that, if a memory operation is followed by a non-memory instruction, the stall is not necessary and they are allowed to proceed back to back, since the second operation does not use the memory controller while advancing through the MEM stage of the pipeline.

We provide an example illustrating a runtime situation for this processor while it encounters the bug described, and discuss how the field-repairable control logic components intervene to bypass it without generating any erroneous result. In the situation, illustrated in Figure 7.7, the program that is being run includes a sequence containing a store and a load instructions back to back, a scenario that triggers the bug described. The FRCL matching logic for the processor contains four entries describing all possible combinations of two memory instructions in the ID and EX stages of the pipeline. For instance, the first entry matches two valid memory read instructions in the ID and EX stages of the pipeline. The second entry matches a store in EX followed by a load in ID: this is the specific entry triggered during the program execution for the example. When the match is triggered, first the pipeline is flushed, then the recovery controller restarts execution in degraded mode at the first uncommitted instruction (in this case an *add* instruction that had reached the memory stage of the pipeline). During recovery, only one instruction can be in the pipeline at a time, thus the *add* would complete alone in five cycles, then the store instruction would execute alone, and finally the load instruction would enter the pipeline. This allows for plenty of extra cycles to complete each memory access required. After a few instructions have completed in degraded mode, normal operation is resumed, and the pipeline is refilled. Note that the bug in the example is completely and precisely described by the four patterns in the matcher, thus no false positive matches are produced.

7.3 Design Flow

In this section we describe a design and verification flow that incorporates field-repairable control logic technology. First, we show what modifications to a traditional design process are required to integrate our solution, and then we investigate how to achieve complete formal verification of the degraded mode of operation. Then we move on to overview techniques for the selection of the control state signals, including an automatic selection algorithm. Finally, we present some insights on incorporating performance critical execution into an FRCL-protected design.

7.3.1 Overview of the Design Framework

As mentioned in Chapter 2, the verification of complex hardware components such as microprocessors relies today on a variety of formal and simulation-based methods. The deployment of FRCL technology in a typical processor design flow requires the addition of two key steps. The first consists in formally verifying all core functionality of the processor when operating in degraded mode; this is required so that FRCL can guarantee recovery from patched design errors. The second step to be added is the selection of the signals that should become part of the "critical control state". The overall FRCL-enhanced flow is shown in Figure 7.8, where the dark gray boxes show the activities and parts involved in the first step, and the light gray boxes show the same for the second step.

Note that in degraded mode instructions are never interacting with each other, hence the formal verification of this mode of operation is greatly simplified over traditional verification efforts of modern processors. For the most part this task is reduced to the verification of individual functional blocks, which are already heavily addressed by formal verification techniques in today's industry. In addition, it is important to check formally that each instruction in the instruction set architecture can individually complete correctly. As shown in the figure, the system-level verification of the complete processor (including the many performance enhancing features) is performed using traditional methodology deployed by the design house

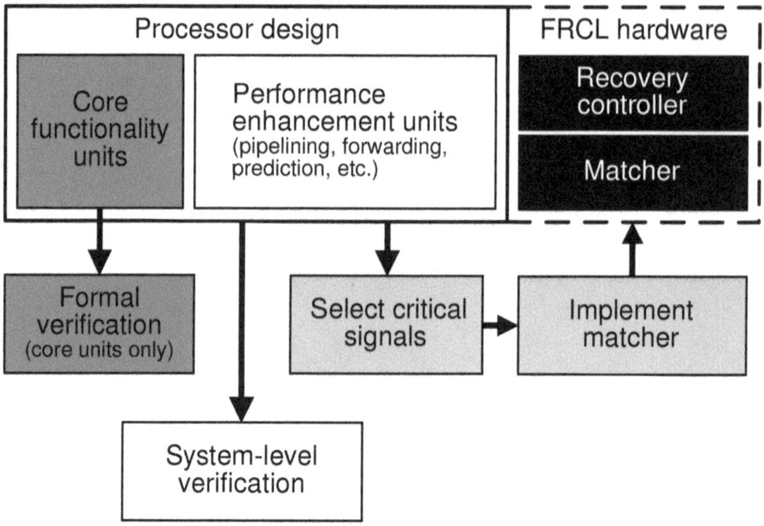

Fig. 7.8 FRCL-enhanced processor design flow. In a FRCL-enhanced design flow, engineers must complete two additional steps over a traditional flow. 1) They must formally verify the functionality of core processor units, active in degraded mode (dark gray boxes) and 2) select control signals to be monitored by the state matcher (light gray boxes). Matcher and recovery controller are then incorporated into the final design shipped to the end customer (black boxes).

before the introduction of FRCL, typically a mix of random simulation and formal property verification.

The control signal selection task entails identifying those signals that should be monitored by the FRLC matcher. The routing of these signals must be enhanced in the design so that they reach the state matcher, as sketched in Figure 7.3. The number of entries in the matcher affects both total design area and overall performance of the deployed component: a smaller matcher might require compression, thus impacting processor performance through a higher rate of false positives.

In addition, engineers must also design a recovery controller to select among degraded and normal mode of operation. The structure of the controller is dependent on the baseline processor architecture, and in most cases this block can be implemented as a simple finite state machine capable of asserting the pipeline's "flush" signal and overwriting the program controller with an appropriate value to start recovery. Both matcher and recovery controller are included as additional components in the silicon die of the final processor before product release.

7.3.2 Verification Methodology

In addressing the formal verification of the degraded mode, we exploited a series of optimizations made available by the intrinsic nature of this mode of operation. Most of the complex functionality of the processor is naturally disabled in this mode and only one instruction is allowed in the pipeline at any time, greatly reducing the fraction of the design's logic involved in proving each formal property. To this end, it is important to note that it is not necessary to create a new, simplified version of the design, instead, all of the optimizations are achieved either as a direct consequence of the nature of the input stream (including only one instruction) or by disabling the advanced features through a few configuration bits (the same that the recovery controller would disable). As an example, modules such as branch predictors and speculative execution units can be turned off with a variant of the "chicken bits", control bits common to many design developments to control the activation of specific features. On the other hand, control logic responsible for data forwarding, squashing and out-of-order execution would be automatically pruned by formal tools, due to the fact that these blocks are irrelevant when pipelining is not active. Together, these two major ways of pruning the design lead to a degraded mode operation that is sufficiently simplified to be tackled by traditional formal verification.

In our experiments we used Magellan from Synopsys [Syn10b] to verify both our testbed processor designs. Magellan's approach to formal verification was first presented in [HSH$^+$00]: it is a hybrid verification tool that employs several techniques, including formal and directed-random simulation. Since in degraded mode instructions are executed independently, we used Magellan to verify the functionality of each instruction in the ISA, one at a time. For each instruction, we wrote assertions in the Verilog hardware design language to specify the expected result. Constraint blocks fixed the instruction's opcode and function field, while immediate fields and register identifiers were symbolically generated by Magellan to allow for verification of all possible combinations of these values. An example of the assertions verified for an *add* instruction is shown in Figure 7.9. The first module in the figure, *add_valid*, guarantees that only valid instructions of type specified, additions in this case, can be executed. The second checker, *add_forward*, enforces forward progress by requiring the instruction to complete within a set number of clock cycles. Finally, *add_sem* enforces the appropriate semantic for additions by checking that the correct result is written to the proper register file location during the writeback stage. For more complex instructions, such as loads and branches, additional checkers are needed to prove that the execution of the operation in the degraded pipelined machine matches exactly the ISA specification. While we could completely verify the degraded mode for both testbeds that we used in our experimental evaluation (as discussed in Section 7.5), it should be pointed out that neither design could be verified in high-performance mode, because of the much greater complexity involved.

```
//a. RTL checker for add validity
module add_valid ( INST, valid, fail );
  assign fail = valid & (INST['OPCODE] != 'add);
endmodule

//b. RTL checker for ADD forward progress
module add_forward (clock, reset, IR, valid, ...);
  reg [3:0] count_committed_adds; //saturating
  reg [3:0] clk_cnt; //saturating
  assign fail = (clk_cnt == 4'd5) &
             (count_committed_adds == 4'd0);
endmodule

//c. RTL checker for add semantics
module add_sem (clock, reset, add_in_id,
          add_in_wb, write_dest, write_data, ...);
  reg [63:0] read_a_, read_b_; //operands from RF
  //result register ID read in decode stage
  reg [4:0] dest_id;
  //shows that destination is chosen correctly
  assign fail1 = !((!add_in_wb) || ((add_in_wb)
              & (write_dest == dest_id)));
  // shows that addition is performed properly
  assign fail2 = !((!add_in_wb) || ((add_in_wb)
            & (write_data == read_a_ + read_b_)));
endmodule
```

Fig. 7.9 RTL checkers used to verify the correctness of the *add* instruction with Synopsys' Magellan. The formal verification tool proves that the checkers hold under any execution scenario, as constrained by **a.** requiring the presence of only one valid *add* instruction in flight. The two checkers for the instruction enforce **b.** forward progress and **c.** correctness of execution.

7.3.3 Control Signal Selection

A critical aspect of deploying field-repairable control logic is determining which control state signals are to be monitored by the matcher. On one hand, it would be ideal to monitor all sequential elements of a design; however, given the amount of control state in complex pipelines, such approach would be either infeasible or extremely costly. For FRCL to be practical, the set of critical control signals should be just a handful, selected among internal nets of the design; although this limitation could potentially be a source of false positives at runtime. An example of the impact of a poor signal selection is discussed in Section 7.5, where we present a bug, *r31-forward*, used in our experimental evaluation, which describes an incorrect implementation of data forwarding through register 31. In the Alpha ISA register 31 is fixed to the value 0, hence cannot be used as a reference register for data forwarding. If the critical signal set does not include the register fields of the various instructions in execution, it is impossible to repair this bug without triggering all those configurations which require any type of data forwarding, causing an extremely high rate of false positives.

We envision two possible solutions to address this problem. The first is to monitor the destination register indices of all instructions at the EX/MEM and MEM/WB stage boundaries by including them in the critical signal set. The downside of this solution is that the critical signal pool would grow and possibly impact processor's performance. For instance, for our in-order experimental testbed this would correspond to a 30% increase in the number of signals monitored. The alternative solution entails including a comparator asserting when a forwarding on register 31 is detected and adding one additional single bit to the critical control pool – the output of the comparator. Both approaches would eliminate the false positives for the *r31-forward* bug and improve the processor's performance, although the additional overhead in the latter case would be smaller than the former alternative. Therefore, a designer deploying the FRCL approach should keep in mind possible corner cases such as this and select the critical control pool taking into account a broad range of bugs. It would be also very helpful to analyze previous designs and gain a sense of where escaped errors have been exposed in the past.

7.3.4 Automatic Signal Selection

Since critical signal selection is of key importance for FRCL, we have developed a software tool to support designers in this task. The tool considers the register-transfer level (RTL) description of a design and it narrows down the candidate pool for the critical control set. It does so by first automatically excluding poor candidates such as wide buses, frequently related to data transfer, and then ranking the remaining ones by decreasing relevance. The rank is computed by taking into consideration the width of the cone of logic that a signal affects and the number of sub-modules that they are connected to. For example, for the RTL block shown in Figure 7.10, the critical state selection tool marks signal A as data (thus not a good candidate), and signals B and C as control. Among the latter two, B is ranked higher since it drives more signals than C (B drives C plus all the nodes that C drives), suggesting that it is probably a more important control signal. When comparing the manually selected critical signal set with the output of the automatic signal selector tool for our experimental testbeds, we noted an 80% overlap. It should be noted that the manual selection was performed by a designer who had full knowledge of the microarchitecture, while the automatic selection tool was only analyzing the RTL description of the processor design.

In Section 7.5 we present an experiment comparing the performance impact, in terms of *specificity* (precision of the bug detection mechanism), of a range of variants of manual and automatic selection. In particular, we analyzed the average *specificity per signal*, a measure of how much each signal is contributing to the precision of the matcher. Solutions with higher average *specificity per signal* provide higher specificity, which translates into higher performance and less area.

```
module example (A, B, C)
    input [64:0] A;
    input B;
    output C;

        assign C = !B & (A == 64'h0);

endmodule
```

A	Data
B	Control rank 1
C	Control rank 2

Fig. 7.10 Example of automatic control selection for a simple module. Signal A is labeled as data and low ranked because of its bit-width. Signal B is ranked higher than C because it has a larger control set: it drives C and, by transitive property, all signals that C drives.

7.3.5 Performance-Critical Execution

For some systems execution performance may be more critical than correctness. For example, in real-time systems, it is important to guarantee task completion at a predictable time in order to meet scheduling deadlines and, in certain applications, partial correctness may suffice. In streaming video applications, the value of the color of a particular pixel may be less crucial than the jitter of the stream. In these situations, a field-repairable control logic approach, trading off performance for correctness, may be undesirable. For these scenarios we propose having an extra control bit to enable/disable the matcher (Figure 7.11). The enable bit, however, should only be modifiable in the highest privileged mode of the processor operation, to ensure that user code does not exploit design errors for malicious reasons.

7.4 Trusted Hardware Design with Semantic Guardians

Although the field-repairable control logic framework enables effective patching of a variety of complex processor bugs in the field, it can only do so with patterns generated offline by a design support team. The most critical consequence of this gap between error discovery and patch availability is the potential for a malicious user to exploit the bug to attack an unprotected system. To prevent this issue and enable the design of truly trusted hardware, we proposed to extend the FRCL solution with a technology called *semantic guardians* [WB07]. A semantic guardian comprises a circuit, generated automatically at design time, which encodes all those design configurations which have not been verified, and thus have the potential to hide a design bug. Thus, the circuit is created at the end of the design process based on the valida-

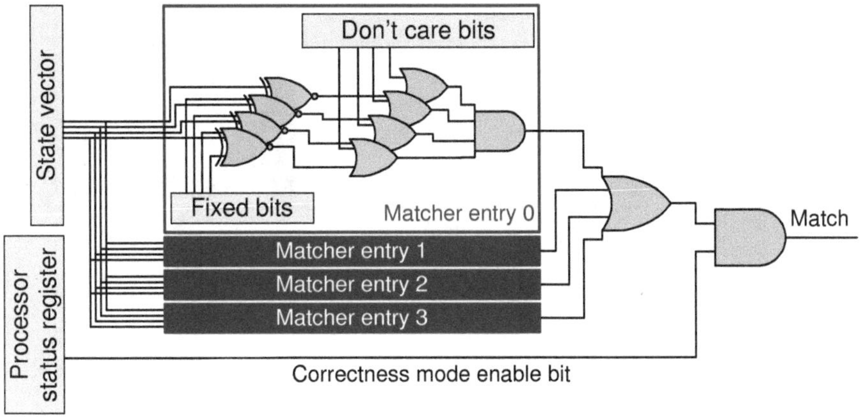

Fig. 7.11 A state matcher for performance-critical execution. Matcher activation can be bypassed by asserting or de-asserting a dedicated bit in the processor status register.

tion effort, and it encodes all insufficiently validated configurations of the processor control state. At runtime, the semantic guardian, similarly to a FRCL matcher, monitors the critical control signals, forcing the design to switch to degraded mode each time an unverified configuration is encountered. Thus, with this solution the processor operates always in one of two state sets: either states validated at design time (when it is in high-performance mode), or states validated formally (when it is in degraded mode). As a result, the chance of incorrect processor operation and the security risks associated with it are eliminated.

The design flow we envision for our solution is shown in Figure 7.12 and it is based on traditional design and verification flows as the FRCL methodology discussed in Section 7.3. We require that the design team formally verifies the units that implement the key functionality of the device, *i.e.,* the *degraded mode* operation. As it is the case for FRCL, in degraded mode the device must be capable of performing all its functionality, however, it can operate at a baseline performance. In the case of microprocessor designs, most design complexity is brought in by performance-enhancing units, thus, this requirement leads to selecting only a few and simple units, leading to a system that can be tackled by modern formal verification tools.

The semantic guardian solution, however, differs from the FRCL framework in its requirements for all remaining units: the goal now is to verify that the most typical and frequently occurring operation scenarios are designed correctly. Indeed, the key aspect of validation in a semantic guardian flow is to provide guarantees that the system can operate correctly in "common" operations. These are typically operations that occur frequently in the system and that also arise most frequently in validation – indeed they are the easiest to reproduce in validation precisely because they are common. The reasoning behind it is that the semantic guardian will trigger the degraded mode every time it encounters a configuration not validated at design

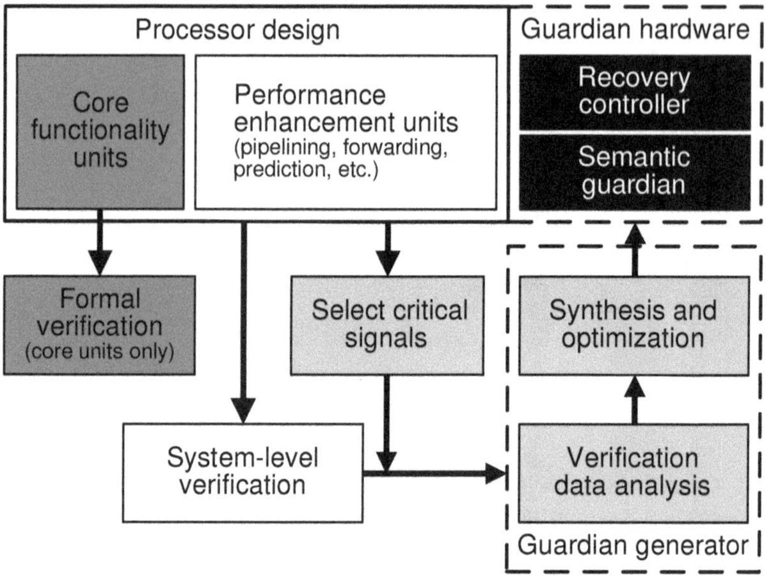

Fig. 7.12 Trusted hardware design flow. The degraded mode is verified thoroughly formally, while the high-performance mode is validated focusing first on the most common functionality. A semantic guardian is then automatically generated based on the behavior that has been validated and manufactured with the design. The guardian, together with a recovery controller, switches the design into the degraded mode whenever an unverified scenario is observed at runtime.

time. Therefore, validating common situations guarantees that the degraded mode will not be triggered too frequently leading to a perceivable performance impact.

An important step in the design of a semantic guardian is the identification of the signals representing the critical internal state of the design. This process can be handled similarly to the corresponding one in FRCL (see Section 7.3.3). However, in contrast with the previous technique, the values of the critical control signals are tracked continuously by a guardian generator during system-level validation. The configurations observed by the generator are labeled as *trusted*, since it is assumed that the verification team has validated the behavior of the device for those states. All others are considered *untrusted*. Afterwards the validation phase is completed, the generator creates a combinational logic block, the semantic guardian, which flags at runtime all untrusted configurations with respect to the critical control signals. The generator also generates automatically a recovery controller, connected to the output of the semantic guardian. The controller has the responsibility of switching the design into degraded mode whenever the guardian flags an untrusted state. It does so by disabling all design's blocks except for the core functionality units. Figure 7.13 shows an example of this solution for a simple pipelined processor. In the figure, the guardian monitors the critical state, and, when an untrusted state is encountered, it signals the recovery controller. The controller flushes all unfinished

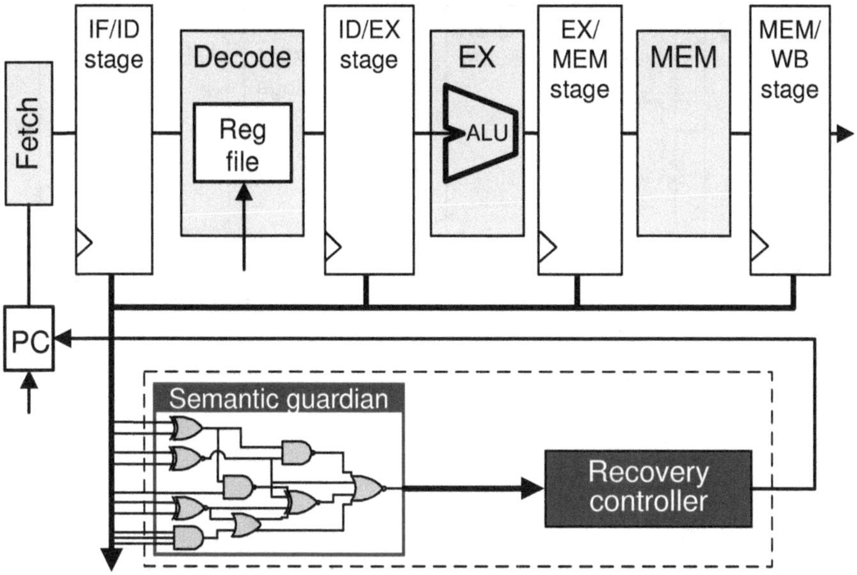

Fig. 7.13 Semantic guardian design for a pipelined processor. The semantic guardian is a combinational logic block connected to the critical control signals in the design and to a recovery controller circuit. When an untrusted state is observed by the guardian at runtime, the recovery controller flushes the pipeline and switches the processor into a degraded mode of operation, guaranteed to produce correct results while making forward progress. Once the untrusted situation is bypassed, the device transitions back to high-performance mode.

instructions and restarts execution in degraded mode from the first uncommitted operation. When the untrusted situation is bypassed, the recovery controller restores the normal high-performance mode of operation.

To optimize area and propagation delay of a semantic guardian block, we propose a range of heuristic optimizations. First, we synthesize both the untrusted configurations set and its complement and select the smaller of the two circuits. Then, if any of the untrusted configurations can be proven unreachable by a formal verification tool, we can leverage this information as a don't care set in optimizing the guardian design. Additionally, designers might choose to trade off the specificity, or accuracy, of the semantic guardian for better circuit parameters. In this case, some of the trusted states might be *re-labeled* as untrusted and included in the corresponding set. This design choice has the potential to increase the effectiveness of the synthesis optimization algorithm, and therefore generate a smaller and faster semantic guardian. On the other hand, it may also trigger false positives in the guardian detection mechanism, hence, it is important to take into account the frequency of occurrence of the re-labeled states.

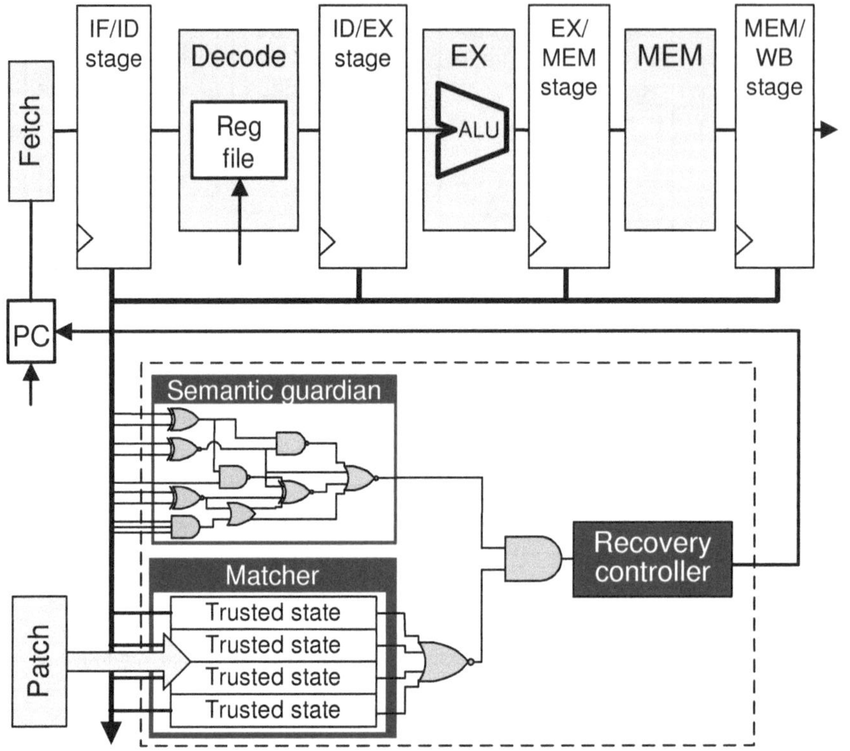

Fig. 7.14 Semantic guardian and patching hardware operating in synergy. A patch extending the set of trusted configurations can be uploaded to the matcher memory after a system is deployed in the field. Afterwards, a match in the semantic guardian can be overwritten by the matcher as a trusted state, avoiding the transition to degraded mode.

7.4.1 Combining Semantic Guardians and Hardware Patching

The semantic guardian solution presented above has the key advantage that it enables trusted hardware execution for software applications running on hardware processors. However, it also has one noticeable drawback – its rigidity. Once the combinational circuit of the guardian is synthesized and manufactured, it cannot be altered, even if, over time, additional verification efforts reduce the pool of untrusted states. This aspect may impact overall processor performance, particularly if the guardian is generated after a short validation effort. To address this limitation we suggest augmenting a guardian with FRCL-style programmable matchers. Since semantic guardians ensure that the system is always operating in a verified state, the patching mechanism in this case is used to overrule the guardian's decision for selected configurations. The configurations of choice would be those that have been proven correct after design's release, resulting in an overall performance boost.

This scenario is illustrated in Figure 7.14: the hardware designer continues device validation after product release. If and when an untrusted configuration becomes sufficiently validated, it is encoded and uploaded via a patch to the matcher memory at runtime. At runtime, if that configuration is ever flagged by the guardian, the patching mechanism overwrites the decision, preventing the transition to degraded mode. Therefore, this approach allows the design team to expand the set of trusted configurations over time and, consequently, improve the performance of the design, even after the device has been manufactured and shipped to end customers.

7.5 Experimental Evaluation

In this section we detail two prototype systems with field-repairable control logic and semantic guardian support. Using simulation-based analysis, we examine the error detection accuracy of FRCL for a number of design error scenarios and a range of state matcher storage sizes. We also examine different criteria for selecting the control state, including an automatic selection heuristic outlined in Section 7.3.4. In addition, we analyze the area costs of adding this support to simple microprocessors and the performance impact of degraded mode execution to determine the extent of error recovery that can be tolerated. Finally, we investigate the semantic guardian methodology outlined in Section 7.4, evaluating several guardian area/delay optimization approaches.

7.5.1 Experimental Framework

To gauge the benefits and costs of field-repairable control logic, we implemented the solution on two prototype processors. Although experimental in nature, these processors have been already deployed and verified in several research projects. While these processor pipelines do not have the complexity of a commercial offering, they are non-trivial, robust designs that can provide a realistic basis to evaluate the field-repairable control logic solution. For our experiments we implemented two variants of the state matcher, one with four and and one with eight entries, and integrated them into the two baseline architectures.

The first design is a 5-stage in-order pipeline implementing a subset of Alpha ISA with 64-bit address/data words and 32-bit instructions. The pipeline a simple global branch predictor, features forwarding from MEM and WB stages to ALU and resolves branches in the EX stage. In our area evaluations we looked at two different cache sizes: small 256-byte direct mapped instruction and data caches and larger 64KB caches. These were used to give us a better estimate of the silicon footprint of our solution in different processor cores. Furthermore, for this design, we hand-picked 26 control bits to be monitored by the matcher. The set of signals governs operation of different logic blocks in the pipeline (datapath, forwarding, stalling,

Table 7.1 In-order processor pipeline: control state bits monitored by the state matcher.

Name	Number of bits	Pipeline stage
IF_valid	1	Fetch
ID_valid	1	Decode
EX_valid	1	Execute
MEM_valid	1	Memory
WB_valid	1	Writeback
ID_rd_mem	1	Decode
ID_wr_mem	1	Decode
ID_cond_br	1	Decode
ID_uncond_br	1	Decode
EX_rd_mem	1	Execute
EX_wr_mem	1	Execute
EX_cond_br	1	Execute
EX_uncond_br	1	Execute
EX_ALU_function	5	Execute
EX_br_taken	1	Execute
EX_hazard_source_a	2	Execute
EX_hazard_source_b	2	Execute
MEM_br_taken	1	Memory
MEM_bus_command	2	Memory
Total	**26**	

etc.) and were selected through a two-step process: we first analyzed a variety of escaped bugs reported in commercial microprocessor errata, and then selected those control signals that would have been good indicators of those bugs. This analysis relies on the assumption that past escaped bugs provide a good gauge of future escapes. In addition, in making the selection, we were careful to choose signals which encoded critical control situations in compact ways: for instance we chose not to monitor the indices of source and destination registers of each instruction (which requires several bits), but instead simply track the occurrence of data forwarding (only a few bits). To limit the monitoring overhead, we also chose not to observe any of the instruction opcode bits that are marched down each pipeline stage. As detailed in Table 7.1, the majority of the critical control signals were drawn from the ID and EX stages of the pipeline, where most of the control logic decisions occur. For example, in the ID stage we selected some of the output bits of the decoder, which represent in compact form what type of operation must be executed, and in the EX stage we selected the ALU control signals. Although this choice potentially limited our capability to recognize a buggy state before instructions are decoded in the ID stage, it also allowed us to reduce the number of critical control bits to be monitored. Note also that, while we did not to modify the original design in any way, it is possible to enhance the accuracy of error detection with minimal additional logic. For example, we did not modify the design to provide an effective

Table 7.2 Superscalar out-of-order processor: control state bits monitored by the matcher.

Name	Number of bits	Pipeline module
ROB_head_commit	2	Re-order buffer
ROB_head_store_address	1	Re-order buffer
ROB_head_store_data	1	Re-order buffer
ROB_head_load	1	Re-order buffer
ROB_full	1	Re-order buffer
RS_full	1	Reservation stations
RS_complete	2	Reservation stations
RS_br_miss	2	Reservation stations
Issue_branch	2	Rename
Issue	2	Rename
Total	**16**	

solution to overcome the *r31-forward* bug discussed in Section 7.3.3. We also did not extend the design with additional pipeline latches to propagate extra information on the instruction being executed through the pipeline: this could be done to capture the specifics of an instruction leading to a bug, and, consequently, improve the precision of the matcher.

The second processor is a much larger and complex out-of-order 2-way superscalar pipeline, implementing the same ISA. The core uses Tomasulo's algorithm with register renaming to re-order instruction execution. The design has four reservation stations for each of the functional units and a 32-entry re-order buffer (ROB) to hold speculative results. For performance reasons this pipeline featured a global branch predictor with branch target buffer. Pipeline flushing for a branch misprediction occurs when the branch reaches the head of the ROB. Memory operations are performed also when a memory instruction reaches the head of the ROB, with a store operation requiring two cycles. The re-order buffer can retire two instructions at a time, unless one is a memory operation or a mispredicted branch. Similarly to the in-order design, we have augmented the out-of-order pipeline with either 256-byte direct mapped i- and d-caches or with 64KB caches for area evaluation. The signals selected for the critical control pool include signals from the retirement logic in the ROB, as well as control bits from the reservation stations and the renaming logic as reported in Table 7.2.

Similarly to the in-order design, to minimize the number of observed signals, no opcodes and instruction addresses were monitored. The matcher developed for this design was capable of correctly matching scenarios involving branch misprediction, memory operations, as well as corner cases of operation in the ROB and reservation stations (*e.g.,* a case when ROB fills up and the front-end should be stalled). Again, a larger set of signals could be used to gather more information about the state of the machine, however, for this design, the benefit would consist only in a shorter recovery time by recognizing problems earlier on. On the other hand, the ability to identify erroneous configurations accurately would not be significantly improved, since errors can still be detected when instructions reach the head of the ROB.

Both processor prototypes were specified in synthesizable RTL Verilog, and then synthesized for minimum delay using Synopsys Design Compiler [Syn10a]. This produced a structural Verilog specification of the processor implemented on Artisan standard logic cells in a TSMC $0.18\mu m$ fabrication technology. For performance analysis, we ran a set of 28 microbenchmark programs, designed to fully exercise the processor while providing small code footprints. These programs included branching logic and memory interface tests, recursive computation, sorting and mathematical programs, including integer matrix multiplication and emulation of floating point computation. In addition, we ran both designs for 100,000 cycles with a dynamically guided constrained-random test generator [WBA07] to verify correctness of operation, as well as to provide a diverse stream of instruction combinations. Finally the core functionality of both systems was formally verified with Synopsys Magellan [Syn10b].

7.5.2 Design Defects

To evaluate the performance of the field-repairable control logic solution, we equipped the baseline designs with a matcher block, manually inserted a variety of bugs into our designs, downloaded the appropriate patch to the matcher, and then examined their overall performance and correctness of execution. In crafting the design flaws to be introduced, we strove to emulate bugs reported in commercial microprocessor errata documents covering all levels of the design hierarchy. Usually, high-level bugs were the result of bad interactions between instructions in flight. For example, *opA-forward-wb* breaks forwarding from the WB stage on one operand, and *2-branch-ops* prevents two consecutive branch instructions from being processed correctly under rare circumstances. Medium-level bugs introduced incorrect handling of instruction computations, such as *store-mem-op*, which causes store operations to fail. Low-level bugs were highly-specific scenarios in which an instruction would fail. For example, *r31-forward* is a bug causing forwarding on register 31 to be performed incorrectly. Finally, the multi-bugs are a combination of errors, forcing the state matcher to recognize larger collections of bug configurations. For instance, *multi-all* is a design variant including all bugs that we introduced. A summary of the bugs introduced in both designs is provided in the Table 7.3. It can be noted that, even for our relatively simple experimental designs, some bugs require a unique and rare combination of events to occur in order to manifest and be detected by the matcher.

Table 7.3 Bugs introduced in the experimental processor pipelines.

Bug name	Description of the error
In-order pipeline	
2-mem-ops	Two consecutive memory operations fail
opA-forward-wb	Incorrect forwarding from WB stage on operand A
opA-forward-conf	Incorrect hazard resolution on operand A
2-branch-ops	Two consecutive taken branches fail
store-mem-op	Store followed by another memory operation fails
load-branch	A conditional branch depending on a preceding load fails
mult-branch	A branch following a multiply instruction fails
mult-depend	Multiply followed by a dependent instruction fails
r31-forward	Forwarding on register 31 is done incorrectly
multi-1	*2-mem-ops + opA-forward-wb + opA-forward-conf + 2-branch-ops*
multi-2	*store-mem-op + load-branch + mult-branch*
multi-3	*mult-depend + r31-forward*
multi-4	*2-branch-ops + mult-branch + load-branch*
Out-of-order pipeline	
rob-full-store	Store operation fails when ROB is full
rob-full-mem	Any memory operation fails when ROB is full
double-retire	Double-issue and double-retirement in the same cycle fails
double-retire-full	Retirement of two instruction fails if two non-branch instructions are added to full ROB at the same time
double-mispred	ROB incorrectly flushes the pipeline if two branches are mispredicted at the same time
rs-flush	Reservation stations do not get flushed on a branch mispredict if rs_full signal is asserted
load-data	Loaded data is not forwarded to dependent instructions in the reservation stations
multi-all	All out-of-order bugs combined

7.5.3 Specificity of the Matcher

The matcher's task is that of identifying when the processor enters a buggy control state, starting the process of switching the system to a degraded mode that provides reliable execution. In this section we study the specificity of the state matcher, that is, its accuracy in entering the degraded mode when and only when an erroneous configuration occurs in the design.

Figures 7.15 and 7.16 plot the specificity of the state matcher for bugs in the in-order and out-of-order processor designs. Recall that the specificity of an error is

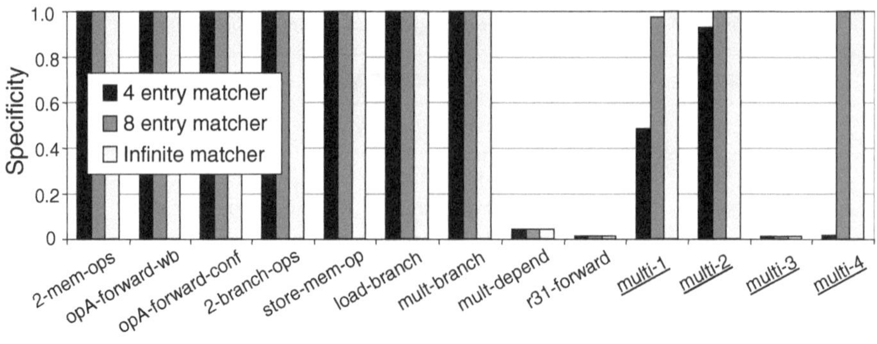

Fig. 7.15 Bug detection specificity in the in-order pipeline. Low specificity, *i.e.*, high false positive rate, can be due to insufficient critical control monitored by the matcher (for instance *mult-depend* and *r31-forward*) or to insufficient matcher size (for instance the 4-entry matcher in bugs *multi-1, multi-2, multi-4*).

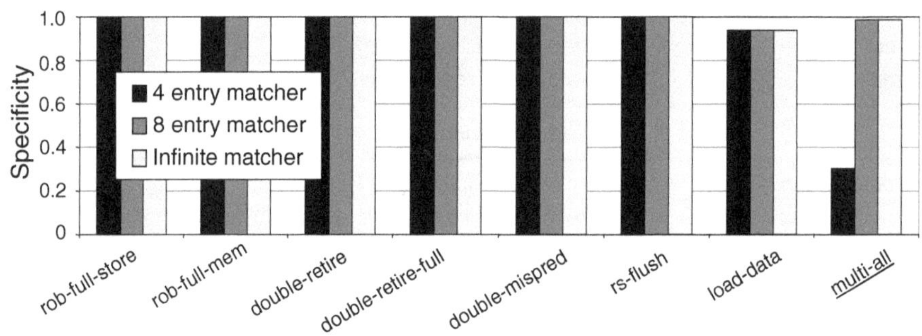

Fig. 7.16 Bug detection specificity in the out-of-order pipeline. Examples of low specificity for this experiment are *load-data*, due to an insufficient number of critical control state bits monitored and *multi-1*, where a 4-entry matcher is not sufficient to store all the bug patterns.

the fraction of recoveries that are due to an actual bug. Thus, if the specificity is 1.0, the state matcher only recovers the machine when the actual bug is encountered. On the other hand, a matcher with low specificity would be overly conservative in its detection and enter the degraded mode more often than necessary. For instance, a specificity of 0.40 indicates that an actual bug was corrected only during 40% of the transitions to degraded mode, while the other 60% were unnecessary. In order to gather a sense of the correlation between specificity and matcher size, we plot our results considering a 4-entry, 8-entry and a infinite-entry, *i.e.*, ideal, matcher. Each of the specificities plotted in the Figures 7.15 and 7.16 were obtained by running the buggy processor design on all the software programs indicated in Section 7.5.1.

It can be noted that for both processors, many of the bugs can be detected and recovered with a specificity of 1.0, even when using the smallest matcher, thus no spurious recoveries were initiated. Some combinations of multiple bugs (*e.g.*, *multi-*

1 and *multi-2*) had low specificities but, upon increasing the matcher size, the specificity reached again 1.0. For these combinations of bugs, a four entry matcher was too small to accurately describe the state space associated with the bugs, but the larger matcher overcame this problem. Finally, for a few of the bugs, *e.g., mult-depend* in Figure 7.15 and *load-data* in Figure 7.16, even an infinite-size state matcher could not reach perfect specificity. For these particular bugs, the lack of specificity was not caused by pressure on the matcher, but rather by insufficient access to critical control information, as described in Section 7.3.3. Thus, in these setups, the matcher had to initiate recovery whenever there was a potential for a bug occurrence, leading to the lower specificities shown.

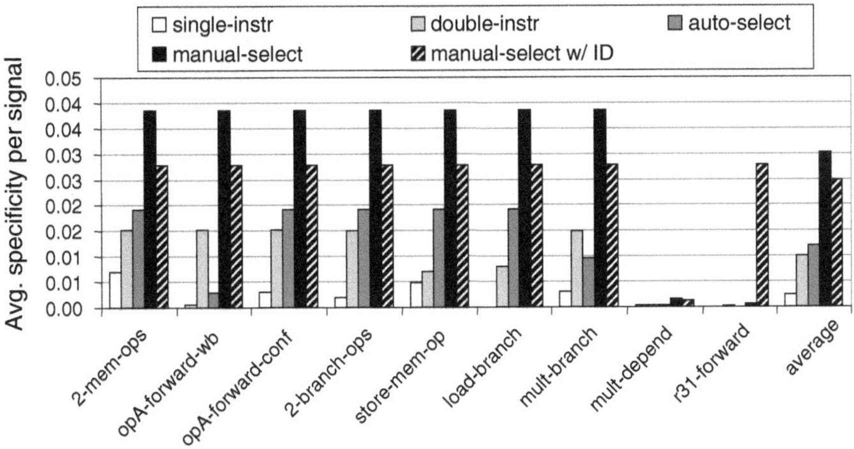

Fig. 7.17 Average specificity per signal for a range of critical signal sets in the FRCL implementation of the in-order pipeline. In this chart we compare the specificity of several signal selection techniques: *single-instr* and *double-instr* monitor one or two instructions in the front-end of the pipeline, respectively; *manual-select* monitors the 26 hand picked signals of Table 7.1 and *manual-select w/ ID* augments it with 10 additional signals. In the *auto-select* approach we used the algorithm of Section 7.3.4 to select critical signals in the processor automatically. After computing the matcher specificity for each of the these techniques, we normalize it to the number of monitored signals, to evaluate the average selection quality. In most cases the *manual-select* solution achieves best specificity at lower cost. However, *auto-select*, based on the automatic selection algorithm of Section 7.3.3 achieves good results with no effort from a designer.

To evaluate the impact of various critical control signal selection policies and compare them to the automatic approach described in Section 7.3.4, we also developed a range of FRCL implementations for the in-order pipeline design, using different set of critical signals. The results of this analysis are shown in Figure 7.17. In the first solution evaluated, *single-instr*, the critical control consists exclusively of the 32-bit instruction being fetched. The second solution, called *double-instr* monitors the instructions in the Fetch and Decode stages (64 instruction bits and 2 valid bits in total). The third configuration (*auto-select*) includes all of the signals selected automatically by our heuristic algorithm from Section 7.3.4 for a total of 52 bits. For this set up the automatic selection algorithm was configured to return all RTL signals with non-zero control rank and bitwidth less than 16. The *manual-select* implementation corresponds exactly to the one from the experiments in Section 7.5.2, that is, all the signals of Table 7.1. The final configuration, *manual-select w/ID* is similar to *manual-select*, but it includes 10 extra signals to monitor the destination registers in pipeline stages MEM and WB.

Matcher sizes for all of the variants contained enough entries to accommodate even the largest patches, so compression was never required. For each design variant, we developed individual patches for the first 9 bugs listed in Table 7.3 (all but the *multi-bugs*). We then measured the *average specificity per signal* for each of them. This metric tracks the specificity divided by the number of signals in the critical control set, providing a measure of which approach has the best performance/area tradeoff. As shown in Figure 7.17, the *manual-select* variant produces the best results for most bugs. The *manual-select w/ ID* solution has better specificity than *manual-select*, but at a higher price. Its main advantage is the good result over *r31-forward*, which is made possible by its tracking destination register indices. Note also that the automatic selection algorithm performs quite well, especially taking into account that this approach does not require any manual effort.

7.5.4 State Matcher Area and Timing Overheads

Implementing a field-repairable control logic solution requires the addition of the critical control matcher logic, that is, the matcher itself and the recovery controller, leading to an area overhead in the final design. Table 7.4 reports the area overhead of a range of FRCL implementations, including a matcher size of 4 and 8 entries included in both the in-order and out-of-order designs and assuming 256B and 64KB instruction and data caches. As shown in the table, the overhead of FRCL is uniformly low. Even the larger state matcher with small pipelines and caches (in-order-256B) results in an overhead of only about 2%. Designs with larger caches and more complex pipelines have even lower overhead. Given the simplicity of our baseline designs, we would expect the overhead for commercial-grade designs to be even lower. Table 7.4 also presents the propagation delays through the matcher block. Note that all solutions have propagation delays that are well below the clock speed. Since FRCL matching is performed concurrently with normal pipeline operation, it does not affect overall design frequency. Finally, note that the matcher for the out-of-order processor is faster because it monitors fewer control signals.

Table 7.4 Area overhead and propagation delay for a range of FRCL implementations for the in-order and out-of-order pipelines. Synthesis was completed targeting the 0.18μm technology node.

	Matcher area (in % of baseline design area)			
	In-order		Out-of-order	
	256B	64KB	256B	64KB
4-entry matcher	1.10%	0.01%	0.34%	0.01%
8-entry matcher	2.20%	0.02%	0.68%	0.02%

	Matcher propagation delay (ns)	
	In-order (clk=11.5ns)	Out-of-order (clk=6.5ns)
4-entry matcher	1.18ns	1.17ns
8-entry matcher	1.43ns	1.21ns

7.5.5 Performance Impact of Degraded Mode

During recovery the processor is switched into degraded mode to execute the next instruction, and then returned to normal operation. Moreover, since in degraded mode only one instruction at a time is allowed into the pipeline, virtually all instruction-level parallelism is lost and program performance will suffer accordingly. Figure 7.18 plots the performance of the in-order and out-of-order processors as a function of increasing recovery frequency. As shown in the graph, for the performance impact to be below 5%, the cost of recovery should not exceed 6 cycles for each 1,000 for the in-order pipeline, and 10 per 1,000 cycles for the out-of-order pipeline. For a more stringent margin of 2% impact, recovery rates should not exceed 2/1,000 and 4/1,000 for the in-order and out-of-order processors, respectively. In addition, recovery has a stronger impact on the in-order pipeline, due to the fact that the out-of-order pipeline can better tolerate the loss of parallelism due to its more capable instruction scheduler. Note that the rates of occurrence mentioned above are considered abnormally high for industrial designs, since even in the pre-silicon verification domain millions of cycles of operation are typically simulated before release. Furthermore, with high-coverage post-silicon validation solutions, such as Reversi (Chapter 3) and Dacota (Chapter 5), most of the commonly occurring errors will be flushed and only very rare corner case bugs will escape. Consequently, in practice, the performance degradation due to FRCL is negligible.

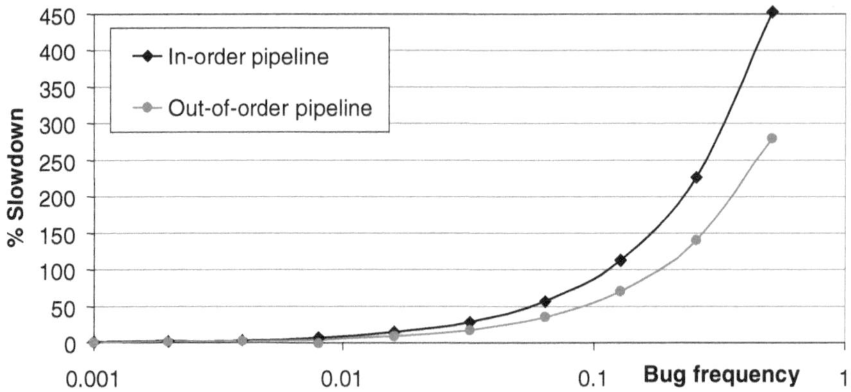

Fig. 7.18 Impact of FRCL recovery on processor performance. Field-repairable control logic technology incurs less than 5% performance impact as long as the frequency of bug occurrence does not exceed 6 per 1,000 cycles in the in-order pipeline and 10 per 1,000 cycles in the out-of-order pipeline.

Finally, Figures 7.19 and 7.20 show the CPI (clock cycles per instruction) of the FRCL-equipped in-order and out-of-order pipelines. The CPI has been normalized to the average CPI achieved when no patch was uploaded on the matcher (hence degraded mode was never triggered). By comparison with Figures 7.15 and 7.16, it can be noted that low specificity often results in increased CPI. However, the worst case, 4 entry matcher on *multi-1* bug, occurs because of an insufficiently sized matcher and not because of the critical control signal selection.

7.5.6 Semantic Guardian Framework Analysis

For our experiments with semantic guardian technology we used the same setup as in the FRCL study, including the the same two processor cores, critical signal sets selected for each pipeline and degraded mode verification with Synopsys' Magellan for all instruction types. For the semantic guardian's synthesis and optimization we used a combinations of several techniques, including different configurations of Espresso [BHMSV84] computing the ON- and OFF-set of the combinational function. We also developed a proprietary heuristic that progressively collapses pairs of states with Hamming distance of 1. Also, if the untrusted set encompassed more than 50% of the total state space, the trusted set was used instead in generating the guardian. Finally, we developed a script which considers the output of Espresso or our minimization heuristic and produces a register-transfer level description of the guardian circuit, which is then synthesized with Synopsys' Design Compiler and mapped to TSMC 0.18 and 0.13 μm libraries.

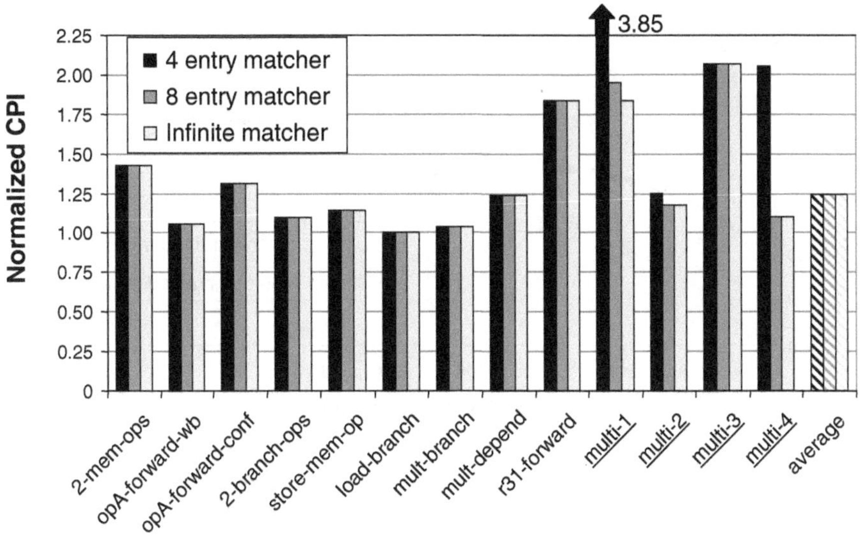

Fig. 7.19 Normalized average CPI of buggy in-order pipeline designs equipped with FRCL. Performance overhead is introduced by periodic transitioning to degraded mode. In the rightmost bars we show the average CPI increase across all individual bugs from *2-mem-ops* to *r31-forward* but excluding multi-bugs to avoid double counting. As the chart indicates, the performance of FRCL-augmented processors depends both on matcher specificity and the frequency of occurrence of degraded mode.

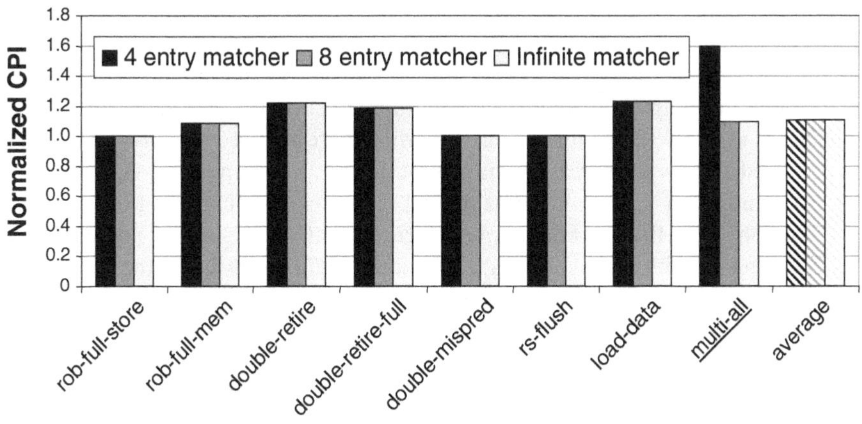

Fig. 7.20 Normalized average CPI of buggy out-of-order pipeline designs equipped with FRCL. The graph reports the performance overhead introduced by FRCL through degraded mode operation, over a range of buggy design variants. In the rightmost bars we show the average CPI increase across all individual bugs from *rob-full-store* to *load-data* but excluding *multi-all* to avoid double counting. Note that the performance overhead incurred by FRCL augmented processors is dependent both on the specificity of the matcher and the frequency of degraded mode operation.

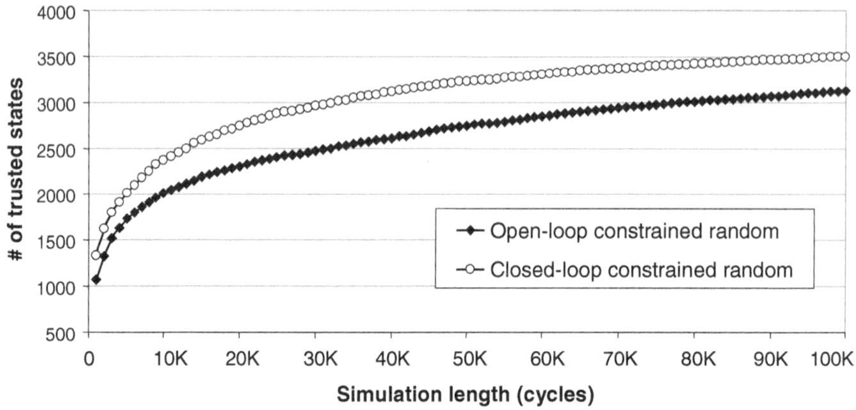

Fig. 7.21 Trusted states vs. simulation effort for semantic guardians. By simulating and validating an increasing number of cycles, more system's states can be labeled as trusted. Moreover, a dynamically guided, closed-loop validation technique performs better then a simple constrained-random approach, reaching higher coverage faster. With a larger number of trusted states the guardian will trigger degraded mode less frequently, leading to higher overall performance.

In our first experiment we calculated the number of distinct critical signal values combinations (trusted states) observed during the validation of the high-performance mode. As input stimuli during this effort, we used both an open-loop constrained-random generator and a dynamically guided, closed-loop, constrained-random generator [WBA07]. In our framework, a state was considered trusted if it was observed at least once during validation, since the signals we selected thoroughly describe the behavior of the processor control state machine. However, users of this solution may choose to require multiple occurrence of a state before it is considered trusted. The results of this experiment are shown in Figure 7.21. It can be observed that, with increasing simulation length, *i.e.,* increasing validation effort, the number of states labeled as trusted increases, leveling off at the far right of the graph. Note that we could not perform formal verification of this complete design because of its complexity and, therefore, could not estimate the fraction of the overall reachable state space explored in simulation.

As a second experiment we analyzed various compression techniques to achieve best area-delay parameters in the semantic guardian combinational logic. The results of this study are presented in Table 7.5. The columns list the compression technique, number of bits in the ON and don't care sets for the Boolean function to be implemented, area in μm^2 and propagation delay through the semantic guardian in *ns* for two technology nodes. The Espresso+ and Espresso- techniques indicate that the circuit was obtained optimizing either the ON-set or the OFF-set of the logic function with Espresso. Compaction is our heuristic optimization technique that progressively collapses pairs of states with Hamming distance of 1, and the last two rows are combination solutions. The top half of the table shows results with no additional optimizations by Design Compiler, while the bottom half shows the best circuit that we were able to synthesize with the most narrow timing and area constraints set in Design Compiler. The best results in both sections are highlighted in bold. Note how most techniques that we analyzed contribute optimal results for at least some metric and the additional Design Compiler optimization always improves propagation delays.

Table 7.5 Semantic guardian design: area and delay achieved with a range of synthesis optimization techniques.

Optimization method	#ON-set	#DC-set	TSMC 0.18 μm #area μm^2	TSMC 0.18 μm #delay ns	TSMC 0.13 μm #area μm^2	TSMC 0.13 μm #delay ns
Without additional Design Compiler optimization						
No optimization	52,104	0	18,600	4.07	9,900	3.78
Espresso+	2,455	11,091	19,400	2.79	10,100	**2.42**
Espresso-	10,586	1,556	26,000	4.13	14,400	3.24
Compaction	18,137	843	**12,100**	**2.76**	**6,200**	2.47
Comp & Espr+	2,449	11,097	19,600	2.78	10,100	2.46
Comp & Espr-	10,586	1,556	25,200	4.44	14,300	3.23
With additional Design Compiler optimization						
No optimization	52,104	0	26,800	1.55	16,800	1.28
Espresso+	2,455	11,091	21,200	1.14	17,100	**0.93**
Espresso-	10,586	1,556	32,400	1.59	26,400	1.27
Compaction	18,137	843	**17,300**	1.19	**14,300**	1.01
Comp & Espr+	2,449	11,097	24,600	**1.08**	16,000	0.96
Comp & Espr-	10,586	1,556	31,600	1.56	26,900	1.30

In particular, we found that the more stringent the area and timing constraints, the higher the priority Design Compiler gives to delay optimizations, producing a significantly faster guardian. We also noted that Espresso generating the ON-set (Espresso+) or Espresso in combination with our compaction approach (Comp & Espr+) generated the circuit solutions with the smallest delay, at a tolerable area penalty. For comparison, the area for the in-order core, excluding caches, in 0.18 μm technology is 500,000 μm^2. Thus, with Design Compiler optimization, the best guardian design can incur as little as 3.5% area overhead.

In addition, we investigated the possibility of generating a smaller and faster circuit by re-labeling some of the trusted states as untrusted. Note that in this experiment, the semantic guardian is generated from the trusted set, therefore, when trusted but rarely occurring states are removed from the set, the guardian becomes smaller. This effect is amplified due to better compressibility of the set when rare states are removed. In other words, Espresso and our compaction heuristic were ca-

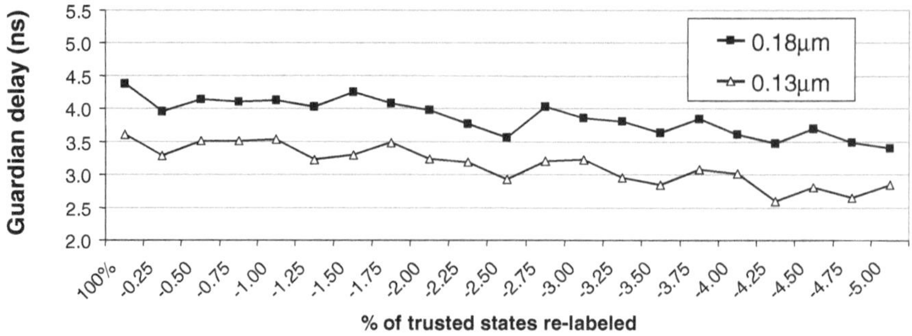

Fig. 7.22 Impact of trusted state re-labeling on semantic guardians circuit delay. Each trusted state observed during validation is labeled by its observation frequency (number of occurrences divided by the total number of simulation cycles). The initial data point is generated by considering all trusted states. Each additional data point is obtained by reducing the overall observation frequency in steps of 0.25% and generating new guardians leading to improved propagation delays.

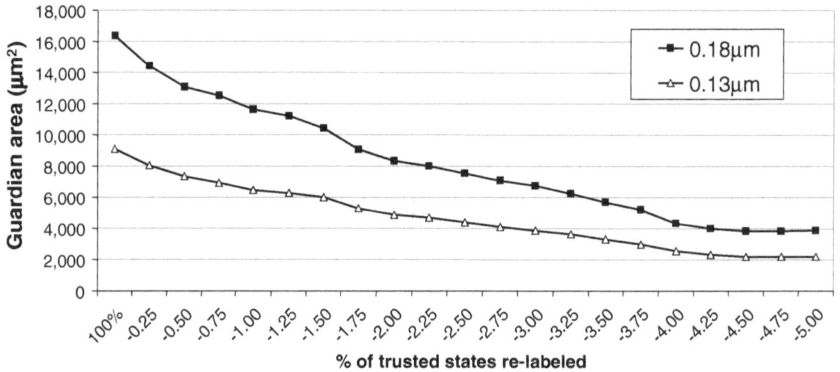

Fig. 7.23 Impact of trusted state re-labeling on semantic guardians circuit area. The plots shows the impact of reducing the pool of trusted states on the semantic guardian area. Each data point is obtained by clustering the observed states by observation frequency and incrementally removing states corresponding to a 0.25% observation frequency reduction. The removal of rarely occurring states leads to better area optimizations in the semantic guardian's logic.

pable of achieving better optimizations without these rare states. It is also possible that, in addition, when these states were removed, more don't care combinations became available, enabling further simplification of the guardian design. For this experiment, whose results are plotted in Figures 7.22 and 7.23, each state observed in validation is labeled by its observation frequency, that is, the number of times the state has been recorded, divided by the total number of simulation cycles. The baseline matcher includes all trusted states and hence its cumulative observation frequency is equal to 100%. Then we grouped states in clusters ordered by cumulative observation frequency, in steps of 0.25%. Thus, for example, the second data point in the figures is generated using observed states accounting for 99.75% of the total observations. Each of the clusters was then progressively removed, one at a time, from the trusted pool, and we created a new guardian each time. All guardians where synthesized with the same settings and without timing or area constraints. Figures 7.22 and 7.23 indicate on the X-axis the trusted-set size reduction in percent. It can be observed from the diagrams that there is a definite reduction in area for smaller trusted sets, enabling up to a 4 times area reduction for just a 5% penalty in observation frequency. This trend can also be observed in the matching delay. The trust re-labeling algorithm allows for a designer to trade off the accuracy of the semantic guardian to improve its area and delay.

7.6 Summary

In this chapter we presented a technology called field-repairable control logic (FRCL): a novel microprocessor patching solution designed to detect erroneous control configurations and recover correct execution via a low-complexity, reliable *degraded mode*. We described an efficient state matching mechanism that can be programmed to detect occurrences of control errors and initiate processor recovery. The technique was experimentally shown to have an area cost of less than 2%. Moreover, with moderately sized matchers, we can ensure highly accurate detection of bug states in nearly all of our experiments. We also showed how the framework can be extended to include semantic guardians - combinational circuits forcing the processor into degraded mode of operation whenever a control state unverified at design time is encountered. Semantic guardians can be deployed separately or they can be complemented by field-repairable control logic to protect processors from unknown escaped errors. Finally, we examined the performance impact on software running on these patched platforms, and found that, if recovery frequency is less than 10 per 1,000 instructions in our out-of-order design, or less than 6 per 1,000 instructions in our in-order testbed, the performance impact is below 5%. Additionally, analysis of semantic guardians demonstrates that these circuits can be subjected to a variety of area/delay optimizations that improve their performance and minimize the silicon footprint. In particular, our experimental evaluation showed that guardians can be optimized to occupy as little as 3.5% of a processor core and can be further optimized using a re-labeling heuristic, trading off guardian performance for silicon area. We also showed that the guardian circuit does not have any impact on the performance of the processor while it is running in a verified state and, as with FRCL, the impact of entering degraded mode is proportional to the frequency of occurrence of untrusted states. Thus, a more thorough verification effort leads to a smaller performance impact after product release.

This chapter makes a strong case for field-repairable control logic and semantic guardians and shows that the approach holds great promise to insure hardware designers and manufacturers against the potential disasters of releasing buggy silicon.

References

[BHMSV84] Robert K. Brayton, Gary D. Hachtel, Curtis T. McMullen, and Alberto L. Sangiovanni-Vincentelli. *Logic Minimization Algorithms for VLSI Synthesis*. Kluwer Academic Publishers, first edition, 1984.

[CBM05] Kai-hui Chang, Valeria Bertacco, and Igor Markov. Simulation-based bug trace minimization with BMC-based refinement. In *ICCAD, Proceedings of the International Conference on Computer Aided Design*, pages 1045–1051, November 2005.

[DDJ06] DDJ Microprocessor Center, 2006. `http://www.x86.org/`.

[HSH+00] Pei-Hsin Ho, Thomas Shiple, Kevin Harer, James Kukula, Robert Damiano, Valeria Bertacco, Jerry Taylor, and Jiang Long. Smart simulation using collaborative formal and simulation engines. In *ICCAD, Proceedings of the International Conference on Computer Aided Design*, pages 120–126, November 2000.

[McC56] Edward J. McCluskey. Minimization of Boolean functions. *Bell Systems Technical Journal*, 6(35):1417–1444, November 1956.

[Syn10a] Synopsys®Inc. *Design Compiler 2010*, 2010. `http://www.synopsys.com/tools/implementation/rtlsynthesis/pages/designcompiler2010-ds.aspx`.

[Syn10b] Synopsys®Inc. *Magellan: Hybrid RTL Formal Verification*, 2010. `http://www.synopsys.com/tools/verification/functionalverification/pages/magellan.aspx`.

[WB07] Ilya Wagner and Valeria Bertacco. Engineering trust with semantic guardians. In *DATE, Proceedings of Design, Automation and Test in Europe Conference*, pages 743–748, April 2007.

[WBA06] Ilya Wagner, Valeria Bertacco, and Todd Austin. Shielding against design flaws with field repairable control logic. In *DAC, Proceedings of the Design Automation Conference*, pages 344–347, July 2006.

[WBA07] Ilya Wagner, Valeria Bertacco, and Todd Austin. Microprocessor verification via feedback-adjusted Markov Models. *IEEE Transactions on Computer-Aided Design of Integrated Circuits and Systems*, 26(6):1126–1138, 2007.

[WBA08] Ilya Wagner, Valeria Bertacco, and Todd Austin. Using field-repairable control logic to correct design errors in microprocessors. *IEEE Transactions on Computer-Aided Design of Integrated Circuits and Systems*, 27(2):380–393, 2008.

Chapter 8
RUNTIME VERIFICATION IN MULTI-CORES

Abstract. In this chapter we shift the focus of our discussion to multi-core processors and issues specific to their runtime verification. In Chapter 4 we overviewed features of modern multi-core designs and described the growing challenge of their verification. As we pointed out, this arduous task is exacerbated by the increasing complexity of the shared memory communication subsystem and the need to verify two of its major system-wide properties: cache coherence and memory consistency. In this chapter we present two runtime solutions designed specifically for this purpose. The first technique, called Dynamic Verification of Memory Consistency (DVMC), designed by Meixner, *et al.* is a checker-based solution, which employs multiple distributed monitors to validate different aspects of communication at runtime. The second solution, Caspar, was developed by us as a patching approach that uses on-die matchers, programmed with patterns describing known bugs, to identify errors. Moreover, to be effective at runtime, both solutions include not only a detection, but also a recovery mechanism, so bugs can be sidestepped and forward progress can be maintained. Thus, as a part of our discussion, we also overview the recovery techniques used in both DVMC and Caspar.

8.1 Dynamic Verification of Memory Consistency

As discussed in Chapter 4, multi-core architectures have become a staple of modern, high-performance processor design, due to their beneficial power/performance characteristics. Since the introduction of the first mainstream dual-core processors, the number of computational elements in these devices have been increasing steadily from generation to generation. While the architecture of individual cores remains relatively simple, compared to the high-end superscalar out-of-order pipelines of the previous decade, on-die interconnect and communication subsystems in these designs are substantially more complex. Furthermore, as the number of cores grows, the interconnect medium starts to move away from relatively unscalable buses to complex interconnect fabric, such as point-to-point or mesh networks. This, in turn,

I. Wagner and V. Bertacco, *Post-Silicon and Runtime Verification for Modern Processors*,
DOI 10.1007/978-1-4419-8034-2_8, © Springer Science+Business Media, LLC 2011

leads to non predictable communication latencies that depend on the network topology and the application's data transfer patterns. As a consequence, the verification of the communication subsystem is becoming more and more important and resource demanding, compared to the validation of individual cores. In particular, this entails verifying such properties as cache coherence and memory consistency, described earlier in Section 4.2. Finally, as design and production timelines keep shrinking, processor manufacturers are forced to release their products well before they are fully verified, exposing end-users to potential escaped errors. To combat the issue, researchers have proposed runtime (dynamic) verification solutions, which augment a baseline design with specialized hardware components, monitoring the state of the system and initiating global recovery if an erroneous operation is detected.

One of the first solutions to address the correctness of operation in multi-core shared memory systems at runtime was developed at Duke University [MS06]. This work, called Dynamic Verification of Memory Consistency (DVMC) is a checker-based solution, which augments individual cores with a set of distributed checkers, dedicated hardware blocks that ascertain that the runtime behavior of the system matches the consistency model specified at design time. If an error is detected, the checkers initiate system-wide recovery, using a checkpointing scheme proposed by Sorin, *et al.* in [SMHW02], called SafetyNet, that we briefly overview later in this chapter. Conceptually, DVMC attempts to verify the three invariants of multi-core operation: *uniprocessor ordering, allowable reordering*, and *cache coherence*. The authors prove that satisfying these three properties is a sufficient condition to guarantee memory consistency. Below we discuss each invariant and then mechanisms deployed to check them in detail.

Uniprocessor ordering. This invariant states that if a shared memory location was accessed by a single core only, load operations issued to this address must return the value of the last store issued by the core to the same address. Thus, private data in any core's cache must remain unchanged, unless modified by the core itself. Runtime verification of this invariant can be enforced by a DIVA-like replay of all memory operations before they commit, to check that values received during replay match those obtained during execution (see Section 6.3 for details of the DIVA technique). To implement this invariant, DVMC requires an extra stage in the processor pipeline, called the verification stage, which re-issues all memory operations just before they retire. Note, however, that in this stage memory operations are issued in commit order, thus correctness of out-of-order execution is checked against the underlying architectural model of memory accesses. The replay of store operations, however, must still be speculative, to prevent these accesses from affecting the architectural state. Thus, the authors augment the system with a *verification cache* (Figure 8.1), which captures write accesses and holds them until they retire.

Allowable reordering is another global property of memory consistency models, which identifies how the order of memory operations can be changed. Consistency models weaker than sequential consistency allow some accesses to bypass preceding operations to improve overall system performance in presence of high-latency

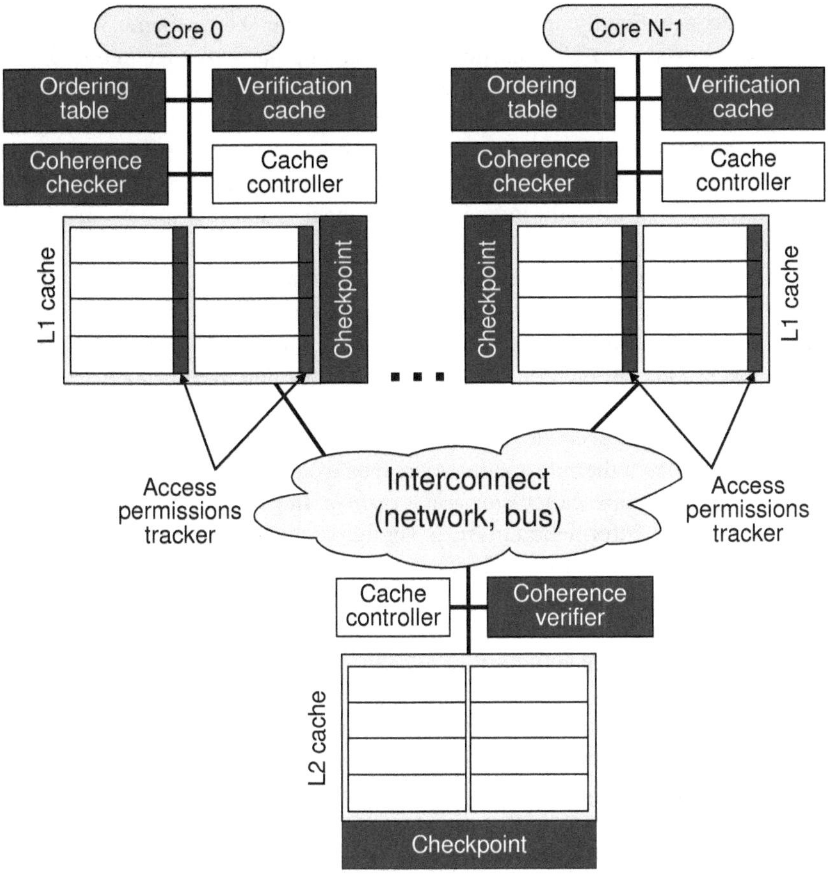

Fig. 8.1 Checker-based approach for Dynamic Verification of Memory Consistency (DVMC). In this runtime solution Meixner, *et al.* validate three invariants of a memory consistency model: *uniprocessor ordering* (checked by verification caches), *allowed reordering* (verified by ordering tables) and *cache coherence* (through the coherence checkers and a shared coherence verifier). The latter requires dedicated storage on each cache like to track access permissions and it entails additional communication overhead. Error recovery is based on a distributed checkpointing system.

operations. For instance, total store ordering, discussed in Chapter 4.2, allows loads to bypass stores to different addresses, while all stores must perform in program order with respect to other stores (hence the name of this model). A few weaker models also exist that relax these rules even further. DVMC can check the enforcement of this invariant with an *ordering table* checker, as shown in Figure 8.1. The pipeline is also slightly modified, augmenting each memory instruction with a sequence number. When a memory operation completes, the ordering table records its number and checks that the operation did not violate the reordering rules of the system's consistency protocol, *i.e.,* operations with higher sequence numbers that must not bypass this access should not have completed yet. In addition, the table keeps track of all outstanding operations to check that all accesses eventually complete.

Cache coherence. Finally, to ensure consistency of the shared memory operation, DVMC enforces system-wide cache coherence. The particular technique for this was described in the early work by the same authors [MS05] and entails checking that each cache line in the system is viewed by all cores as either "Shared" or "Exclusive". The period of time during which a line is in one of these states is called an epoch. Thus, to validate the cache coherence invariant, one must check that accesses are only performed during appropriate epochs, and that epochs in which a line is in Exclusive state in one cache do not time overlap with Exclusive epochs in another cache. To this end, each core is augmented with a *coherence checker* (Figure 8.1), which keeps track of active epochs for each cache line and periodically sends information messages to a memory controller. The messages include such data as the logical time of the beginning and the end of the epoch, as viewed by this particular core, as well as cache block checksums, which enable detection of data corruption, rogue accesses, *etc.* The memory controller is also extended with a table that compares the information messages from individual cores and detects potential epoch overlaps, called *coherence verifier*. In general, Meixner shows that the overhead due to information traffic is negligible, however, note that this system requires synchronization of the coherence checker's clocks, which may become a problem in large systems with non predictable interconnect delays. Moreover, the coherence checker has a somewhat large die area overhead, at least 7% of a node's cache, as estimated by the authors.

It is important to note that the specific implementation of the three invariants is irrelevant in guaranteeing memory consistency. Therefore, individual detectors, for instance the coherence checker, can be designed differently, *e.g.,* for a different area vs. performance tradeoff, without compromising the runtime guarantees of the framework. The same observation holds for our runtime solution, called Caspar, presented in detail in the next section. Since Caspar is not a checker-based technique, but a patching solution, it trades off area overhead of a coherence checker for the requirement to develop error patterns offline.

The final aspect of DVMC that must still be investigated, is the recovery scheme used to ensure forward progress after an error is detected. As described in the original work, recovery from errors that have not altered the architectural state of a system can be as straightforward as a pipeline flush and restart. However, the vast majority of problems in multi-core processors arise not because of flaws at the local memory interfaces, but are due to erroneous interactions among cores. In these cases, by the time the error is detected by one core, system-wide architectural and microarchitectural state may have already been affected, and one must resort to checkpointing mechanisms to restore the last known-correct global state and restart the execution. This was precisely the approach that the authors of the DVMC project took, utilizing a checkpointing solution called SafetyNet [SMHW02], which they had previously developed.

SafetyNet relies on logging data in dedicated storage distributed throughout the system, and on a synchronization subsystem that binds individual cache transactions to invidual epochs. The framework uses a checkpointing clock - a loosely synchronized logical time base, which individual nodes use to mark the beginning of a checkpointing interval, as well as to identify transactions belonging to a certain interval. To this end, an individual coherence transaction (consisting potentially of multiple communication steps) is defined to belong to the epoch during which the previous owner cache received the request and relinquished the ownership of the line. Thus, upon receipt of a release request, the owner is responsible of piggybacking the current epoch's ID onto the response, so that the requester may label the transaction appropriately. Two important consequences of this fact are the apparent atomicity of memory transactions and unique and definite binding of each transition to an individual epoch throughout the system. The changes, caused by these transactions, are maintained in storage logs, associated to each cache in the system. For optimization of storage, SafetyNet features log-on-change policy, recording the changes in local cache data/permissions only if they have been modified since the beginning of the last checkpointing interval. Thus, if a checkpoint needs to be restored, the logs are marshaled in reverse order, reverting the changes and restoring the global state. Potentially, SafetyNet can store multiple checkpoints simultaneously, so that the system can rollback to various points in the past execution. However, for the purposes of runtime verification, a single checkpoint suffices. Among the limitations of SafetyNet are the need for special techniques for replaying I/O operations and the need for synchronization traffic or special checkpointing clock. The former solution has been addressed in such related works as [NMGT06], while the latter is discussed further in this chapter as part the presentation of the checkpointing system used in Caspar.

DVMC was the first system to approach the challenge of runtime verification of multi-cores in an organized and consistent matter: it contained both a distributed detector network and a recovery solution. However, being a checker-based technique, it incurred somewhat large area penalty, since individual blocks had to be "smart" to encode various memory consistency invariants and detect violations thereof. In our discussion of dynamic verification of individual cores in Chapter 6 we presented hardware patching as a viable alternative to checker-based solutions. Thus, we propose to apply the same concept to runtime verification of a processor's cache subsystem and we investigate the benefits and drawbacks of this type of solution in detail in the next sections.

8.2 Caspar: A Multi-core Patching Solution

The remainder of this chapter presents Caspar (CAche Subsystem PAtching and Repair) – a novel patching-based runtime validation solution that was designed specifically for multi-core processors. Caspar incorporates on-die cache-event detectors to identify erroneous situations. Each detector is programmed at system startup with *error-condition patterns* that are compared to a cache line's state each time a transition occurs in a local or shared cache. The patterns themselves are created by the chip manufacturer and distributed to end-customers when new errors have been discovered and debugged. When a memory-related error is detected by Caspar, a recovery and bypass sequence is triggered. The Caspar solution includes also a lightweight checkpointing system to record the state of processor's cores and caches at regular intervals. In case of error, the recovery process begins by restoring the last checkpoint snapshot of the system. Then, it replays the most recent execution by relying on a reduced-complexity operation mode of the system, where only one memory access is performed at a time. This style of execution ensures that there are no interactions among the processor cores, therefore, memory coherence and consistency can be guaranteed. Finally, after the recovery process has been executed for a predefined period of time, the system resumes regular operation.

One of the key advantages of Caspar is its small performance impact in the absence of bugs. On the other hand, when the processor contains errors, the performance of the system depends on the frequency of occurrence of the errors. Moreover, the event detectors in Caspar occupy less than 0.01% of the silicon die area (for the OpenSPARC T1 machine), and the majority of the area overhead is due to the checkpoint storage space. Note that checkpointing can be leveraged for purposes beyond error recovery (*e.g.,* post-silicon debugging), thus amortizing the cost of implementing Caspar.

8.3 Caspar's Design

In this section we overview the structure and operation of Caspar. We assume a generic multi-core architecture, similar to the one in Figure 8.2. As shown in the figure, the processor usually consists of multiple cores, each with a local L1 cache. Controllers manage the caches, satisfying local requests, as well as requests from other cores received through on-chip interconnect. In addition to the local data storage, the cores have access to a larger shared second-level cache. To deploy Caspar in such a system we require the addition of a few hardware modules to the baseline system (darkened blocks in the figure). Note that our solutions is not limited to this architecture and can be used in other multi-core designs with multiple levels of caches, hierarchical or point-to-point interconnect, *etc*. The majority of the area overhead is due to system-wide recovery, which requires storage space as it is implemented through checkpointing. In Caspar we create a checkpoint of each core's state by logging the values of its architectural registers. To checkpoint caches we implement a copy-on-write policy and record state and value of a line upon change. Note that checkpoints in Caspar use local storage, eliminating the need to transfer logged snapshots. In addition to checkpoint storage, Caspar augments each cache with a programmable *cache-event detector* – a specialized circuit that compares transitions experienced by individual lines with a pre-loaded pattern, thereby identifying memory subsystem errors. If a cache transition matches a detector pattern, Caspar alerts the cache controller, which, in turn, signals the second-level cache to initiate recovery and bypass.

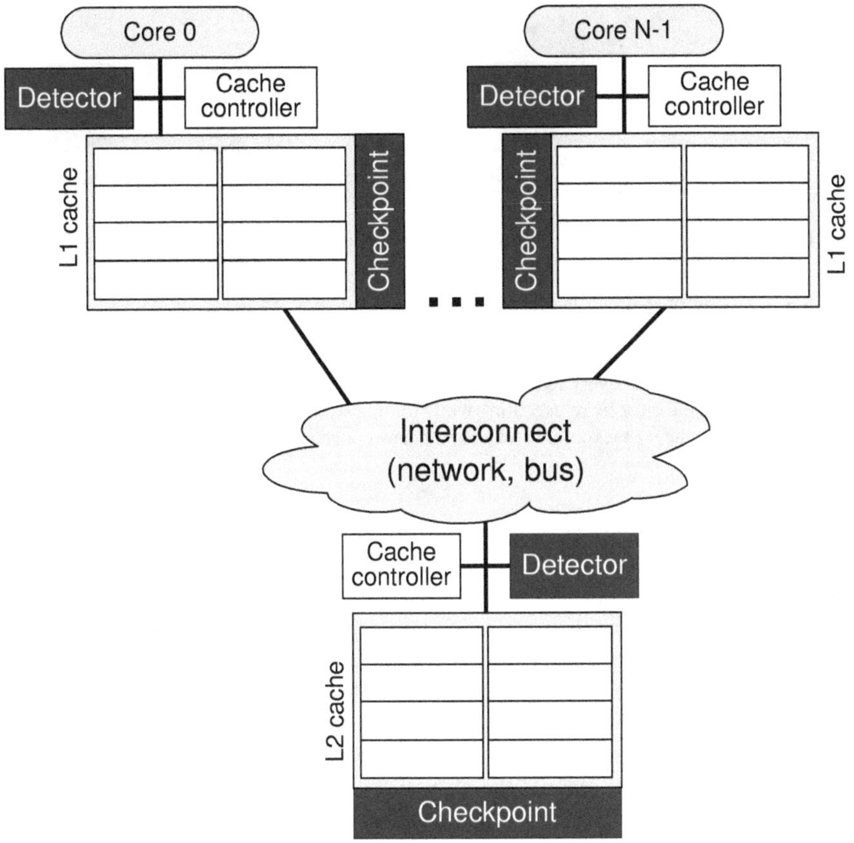

Fig. 8.2 Architecture of a multi-core system augmented for Caspar support. Caspar requires the addition of detectors at each cache storing error-condition patterns and storage space for system state snapshots taken at regular intervals.

Caspar's operation is organized into epochs, as shown in the schematic of Figure 8.3. We assume that Caspar's event detectors are loaded with appropriate patterns describing erroneous events at startup. At the end of each *checkpointing epochs* the system is synchronized and its state is recorded. *Synchronization* in our framework is initiated by one of the cache controllers, level 1 (L1) or level 2 (L2), and forces completion of all cache transactions in flight. During the synchronization phase no new memory accesses are allowed to enter the system, thus all cache lines reach stable protocol states, after which, at the end of the epoch, a new distributed checkpoint is taken. In the second epoch of the example shown in the figure, one of the detector circuits identifies a cache transition corresponding to a known coherence issue and triggers the recovery protocol, which stops program execution and restores the last checkpoint. The L2 cache controller then starts polling individual cores, allowing only one memory access at a time over the entire system, thus ensuring absence of coherence race conditions and enforcing strict ordering of memory accesses. When

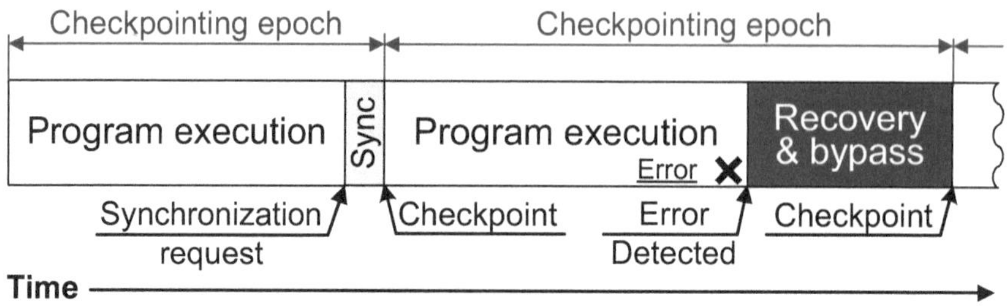

Fig. 8.3 Timeline of a possible Caspar operation sequence. In the example no errors occur during the first epoch, which completes with system synchronization and a new checkpointing snapshot. During the second epoch an error occurs and it is detected by Caspar's cache-event detectors, triggering the recovery mechanism. The previous checkpoint is restored, then the system attempts to bypass the error by re-executing while forcing strict ordering of accesses. After forward progress is made, a new checkpoint is taken and a new epoch commences.

a user-specified number of accesses completes, Caspar stores a new checkpoint, and the cores resume operation at full performance. Note that after completion of the last memory operation in recovery mode, the system is again in a stable state, since during recovery only one memory operation was executed at a time, thus guaranteeing that no memory protocol issues due to core interactions could occur.

As discussed above, Caspar can be decomposed into three major components: the *event detectors*, which identify bugs, the *recovery and bypass mechanism* that allows Caspar to restore the last valid state and circumvent the error, and the *checkpointing system* that periodically records the system's state. In the rest of this section we discuss each of these components and detail their implementation in Caspar.

8.3.1 Detection and Coverage

Caspar is designed specifically for patching errors in the memory subsystem by comparing cache events (cache controller's state machine transitions) to pre-loaded patterns describing known bugs. Thus, Caspar can cover a wide class of functional errors associated with unexpected events that are either handled improperly or not handled by the coherence controller. Moreover, as we demonstrate later, Caspar can be extended to detect and recover from memory consistency errors. However, in its current implementation, Caspar is ineffective against bugs that cannot be represented through local state-machine transitions and are caused by invalid request timing, *etc.* The detection feature of Caspar is implemented with event detectors – small programmable blocks residing at each of the system's caches. Since each coherence transition of a cache line is uniquely defined by a line's present state and the input trigger event, the error patterns in the detectors consist of two fields, matching the state and the event, respectively. Each field is stored as a bit vector, where a 1 in any particular position matches a certain trigger or a state. Note that multiple bits in a pattern can be set simultaneously to encode combinations of erroneous transitions.

The diagram of the hardware event detector is shown in Figure 8.4. During every coherence transition, the line's state and the input event are passed to one-hot encoders, which convert them to bit-vectors describing the transition. The vectors are then separately compared with stored patterns on a bit-by-bit basis. The error detection signal, however, is asserted only if the match occurs in both fields. Caspar detectors can also be extended to contain several entries, each identical to the one in the figure, to perform multiple comparisons in parallel.

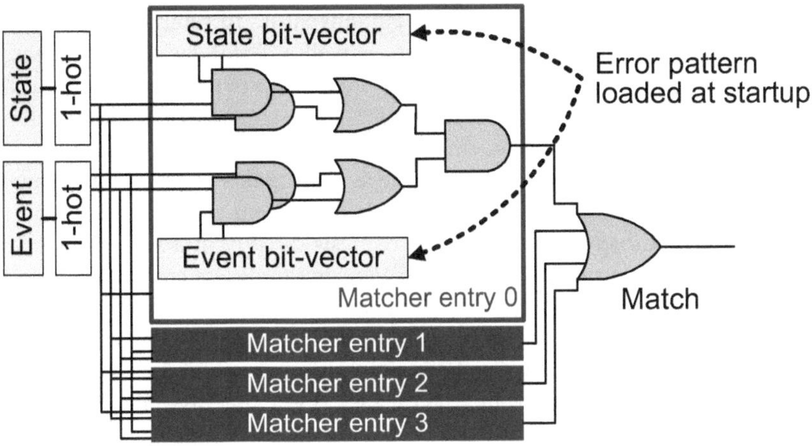

Fig. 8.4 Caspar's hardware event detector. Each bug pattern is partitioned into a state and an event field, stored in two bit-vectors. Each bit set to 1 in a bit vector corresponds to an erroneous state or trigger event. During each cache access, the detector encodes the current transition as two one-hot encoded vectors and compares them with the stored patterns, asserting an error detection signal only if at least one state and event vector pair match.

We now illustrate Caspar's detection with the following short example. Assume that a core is performing a store and the corresponding address line is in the intermediate state "SM" (half-way between "Shared" and "Modified"), when an invalidation message arrives from another core. If the cache's finite state machine was incorrectly implemented, this race could result in a deadlock or coherence violation in a non-Caspar system. However, by programming Caspar with a pattern describing this state ("SM") and trigger (invalidation event), we can detect and avoid this error at runtime.

In addition to coherence errors, the Caspar framework enables patching of more complex and subtle issues related to memory consistency. As it was demonstrated in the DVMC solution, consistency can be ensured if three system-wide invariants are guaranteed: *uniprocessor ordering*, *absence of illegal reorderings* and *cache coherence*. As discussed above, event detectors in Caspar recognize dangerous transitions in the L1 and L2 caches, thus ensuring that the third invariant is upheld. Moreover, as was demonstrated in Section 8.1, the checkers for the former two properties can be implemented efficiently in small logic blocks (non-programmable) that can signal errors similarly to event detectors. With the addition of such checkers the coverage of our approach can be increased to include bugs which manifest themselves as memory consistency violations.

8.3.2 Recovery and Bypass

Once an error in a program's execution is detected, Caspar initiates a system-wide mechanism to recover from the bug and circumvent it. To return the system to a know-correct state Caspar forces all cores, as well as local and shared caches, to restore the last checkpoint. Then the system attempts to bypass the error to ensure forward progress. Since errors in the memory subsystem are triggered by interactions of simultaneous accesses, we designed a bypass system to eliminate such race conditions, at the cost of reduced performance. During this phase of operation, we only allow one outstanding memory access in the system at a time, reconfiguring the L2 cache controller to poll cores for requests directly in a round robin fashion. Note that, in this case, coherence transitions in the system are never suspended in intermediate states, and cache operations are simplified to the level of the underlying high-level coherence protocol. Thus, correctness of the bypass operation is sufficiently simplified that it can be verified using powerful formal techniques. Moreover, since the second-level cache (or memory) explicitly orders each core's memory accesses, consistency in bypass mode can also be ensured.

After predefined number of accesses complete in bypass mode, the second-level cache controller broadcasts a new checkpointing request, so that the stable state reached in bypass is saved, and normal execution may resume. Note that many more accesses are usually performed during an epoch's normal execution than during bypass, therefore, it is possible that the offending sequence of instructions reoccurs after recovery completes, and the error is triggered again. However, in this case the system rolls back only to the most recent checkpoint, reached after the last recovery, thus guaranteeing forward progress so that the error is eventually circumvented. Furthermore, since in flight data and message buffers are drained during recovery, the specific interaction of accesses after bypass is often different than the one that triggered the recovery originally. Thus, it is possible that when normal operation resumes, the execution of the same instruction sequence does not produce the same erroneous event again.

8.3.3 Checkpointing

Checkpointing in Caspar is designed to allow fast and efficient system recovery if an erroneous event is detected. As mentioned above, storage for checkpoints is distributed throughout the system to reside locally with each module whose state is to be recorded. The implementation of a core's checkpointing mechanism is straightforward, since in microprocessor pipelines it is sufficient to store the architectural state in order to enable correct replay. In Caspar we provide checkpoint storage for the cores via shadow registers that contain the architectural state at each snapshot. Caches and memories are too large to be checkpointed in such a way, so for them Caspar adopts a copy-on-write policy, recording the previous value or the coherence state of a cache line only upon its first change after the beginning of a new epoch. To this end, we augment each cache line with two *modification bits* to flag the modification of a line's data and state during the current epoch. We also create a local log organized as a hardware queue that records address, original state and data of altered lines. The queue and the modification bits are accessed concurrently with the main cache operation and are cleared when a new checkpoint is taken. Then, if a line's state is altered during an epoch, we set the corresponding state modification bit and append the line's address and previous state to the log queue. It is important to note that if only the state is modified, then the line's data is not copied to the log, allowing for better utilization of the queue storage space. If the block is subsequently altered again, the new change will not be logged, since we are only recording the information necessary to recreate the system's state at the time of the last checkpoint. To restore the last checkpoint state, Caspar suspends the cores' operation and performs the following two tasks: each core's architectural state is restored from the shadow registers and, simultaneously, we traverse the log queues in reverse order, restoring the state and/or data of the modified cache lines from the logs while resetting the modification bits. When the traversal is completed, the log queues are cleared.

Although cache checkpoints in the system are distributed, their recording must be synchronized, ensuring a consistent system state after recovery. In designing Caspar we decided to avoid specialized physical clocks to synchronize the cores for checkpointing, as required by solutions such as SafetyNet. Such clock networks would require significant routing effort to be implemented in a large multi-core CPU, and incur additional power and area overhead. Instead, Caspar relies on a four-step procedure, as illustrated in Figure 8.5. The synchronization is invoked by either a signal from the watchdog timer in the second-level cache (1.a in the figure) or a message from a cache controller that has exhausted its logging storage space (1.b). Subsequently, the second-level cache controller broadcasts an explicit synchronization message (2) stopping program execution in all cores. Once all cores have acknowledged completion of all pending operations (3), the L2 broadcasts a "checkpoint" message and the state of the device is recorded as the new checkpoint (4).

8.4 Post-silicon Debugging with Caspar

In addition to providing runtime protection from subtle cache coherence and memory consistency bugs, Caspar can be used effectively in a variety of post-silicon debugging and software profiling frameworks. To this end, checkpointing and recovery

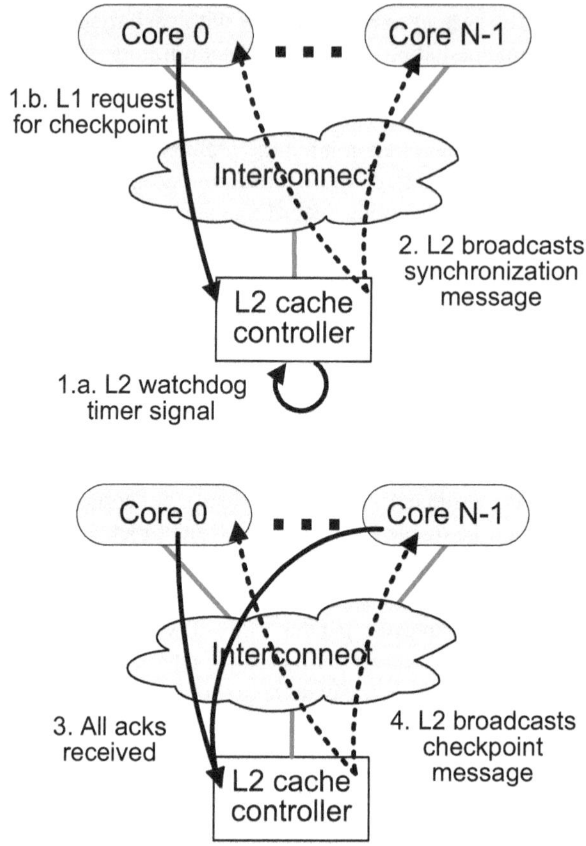

Fig. 8.5 Synchronization for checkpointing snapshot in Caspar. Synchronization requests are initiated either by the watchdog timer in the level 2 cache controller (1.a) or by a local cache controller that exhausted its log space (1.b). Following the request, the L2 cache controller broadcasts a synchronization message to all cores (2), who, in turn, stop issuing new memory accesses, wait for completion of all pending operations and, finally, acknowledge synchronization (3). As the final step, the L2 controller sends a "checkpoint" message triggering the architectural state of the cores to be saved, and future changes in local caches to be logged.

capabilities are disabled, allowing the system to run at full speed. In addition, the detectors can be reconfigured to monitor and count the number of occurrences of each event, or to issue a system trap on a match. In this setup, the patterns loaded in the detectors do not describe erroneous events, but rather transitions that are of particular interest to the validation team. The system can be further enhanced with dynamic pattern replacement to detect complex sequences of memory subsystem events and further aid in the debugging process.

8.5 Experimental Evaluation

To evaluate the performance of the Caspar patching solution, we used a setup similar to the one used to analyze our post-silicon validation system Dacota (see Chapter 5). We simulated the operation of Caspar in a 16-node multi-core processor using the Wisconsin Multifacet GEMS memory subsystem simulator [MSB+05]. The local L1 caches were 64KB each, while the shared L2 cache had a size of 4MB. Cores were interconnected with a mesh network and used the MOESI directory protocol for coherence. Caspar was implemented inside the GEMS memory tester module as a collection of functions to initiate checkpointing, recovery and bypass procedures, and to manage the event detectors, which were inserted into the system's caches and directory controller.

Caspar was tested using a set of twenty benchmarks that included ten traces of SPLASH2 suite [WOT+95] and additional randomly generated stimuli. We sampled the SPLASH2 benchmark traces for 10,000,000 instructions, using the Virtutech Simics simulator [MCE+02]. Random benchmarks, on the other hand, contained 1,000,000 memory accesses and varied in their degree of data sharing and total address space used. Several of these synthetic programs had a small data set and could fit entirely in the L1 caches of individual cores, thus eviction was never invoked. Other random benchmarks has a much larger profile than the side of the level 1 system caches so, throughout their execution, lines were frequently moved between the shared L2 cache and the local caches. Two of the synthetic benchmarks used GEMS' built-in random traffic generators executing the "barrier" and "locks" patterns.

8.5.1 Error Resiliency Analysis

In our first study, we created thirteen different buggy versions of the multi-core processor under test, by injecting a different bug in each variant. The bugs, listed in Table 8.1, affect various modules of the design and were derived from publicly available errata of commercial multi-core processors and then adapted to our evaluation framework. Each bug was associated with a particular state machine transition and was described uniquely by the error patterns loaded in Caspar's event detectors. These errors represent complex corner cases of operation of the L1 caches, the L2 cache and the directory unit, when multiple accesses race for resources, or an unexpected request arrives when the system is in an intermediate coherence state. Memory-related errors such as these can often lead to cache corruption and/or deadlock in real life systems.

For each bug, we calculated the frequency of occurrence by recording the number of times the corresponding state and trigger combinations occurred during normal operation (*i.e.,* with checkpointing and detection disabled) while running all of our testcases. These frequency values are reported in the last column of Table 8.1. We then re-run our experiment with Caspar in place and found that, when the patterns identifying the bugs were loaded into the detectors, all test completed correctly.

Table 8.1 Design bugs injected in our experimental platform with relative frequency of occurrence (occurrences/1,000,000 cycles).

Bug name	Description of the error	Frequency
L1 Store race	L1 transitions from S to M while another cache issues a store	0.32
L1 WBack+Load	L1 evicting dirty data while another cache issues a load	0.72
L1 Owner Evict +Load	L1 owner evicting dirty data while another cache issues a load	0.04
L2 Block+Unblock	L2 blocked, waiting for ack while requester issues unblock	34.3
L2 Block+Load	L2 blocked, waiting on directory while L1 issues a load to the line	0.03
L2 Load race	L2 load while another load arrives	0.12
L2 WB	L2 WB while dirty data arrives	2.72
Dir Block+Unblock	Directory blocked on a load while L2 issues unblock	35.1
Dir Clean WB mem	Directory blocked on WB while clean WB forwarded to mem	12.3
Dir Dirty WB mem	Directory blocked on WB while dirty WB forwarded to mem	21.7
Dir Clean WB L2	Directory blocked on WB while clean WB forwarded to L2	0.04
Dir Dirty WB L2	Directory blocked on WB while dirty WB forwarded to L2	0.09
Dir Store+Unblock	Directory blocked on a store while L2 issues unblock	34.2

8.5.2 Checkpointing Overhead

In our second study we analyzed the overhead of Caspar's checkpointing process. Recall that, to take a checkpointing snapshot, the system must first synchronize, and thus the execution of some memory accesses must be delayed. Consequently, longer epochs result in less frequent synchronization, and less delay impact on the application. On the other hand, due to the copy-on-write policy for cache checkpointing, longer epochs can only be achieved with larger log sizes, which in turn imposes a higher area overhead. We investigate this tradeoff in Figure 8.6, where we plot performance and area penalty for a range of epoch lengths. The area overhead is calculated as the percentage of cache lines that were modified on average during each simulation's epoch, and thus had to be logged. Each data point is computed as an average over the twenty benchmarks in our setup. Individual benchmarks deviated slightly from this average: less than 4% for synthetic random programs and less than 1% for SPLASH2 applications. As the figure indicates, longer epochs lead to better performance when no errors manifest, as in our setup for this evaluation. However, in presence of errors, longer epochs will have a higher performance impact because the number of cycles spent in bypass mode increases – most often due to several recovery events caused to a same bug manifesting far from the beginning of the epoch. Based on these findings, for the subsequent experiments, we set the checkpointing epoch length to 8,000 cycles, which translates to roughly a 2.4% performance impact and a 5.8% area overhead.

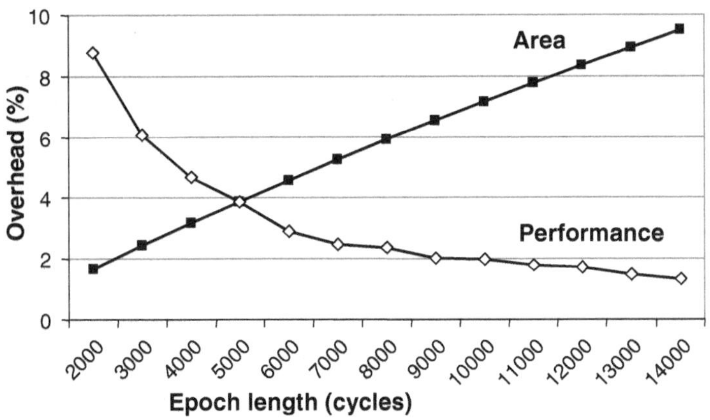

Fig. 8.6 Performance and area of Caspar's checkpointing system. The chart reports performance and area overhead imposed by the checkpointing infrastructure for varying epoch lengths. With longer intervals between checkpoints (*i.e.,* epoch lengths) more cache lines must be logged, leading to higher area overhead. Performance overhead, on the other hand, is smaller for longer epochs, due to the fact that system synchronization is invoked less frequently.

8.5.3 Caspar Recovery Performance

In addition to evaluating the overhead of our solution strictly due to checkpointing, we also investigated the performance overhead entailed by the Caspar detection and recovery mechanisms by executing our twenty benchmarks in presence of different bugs. In this study, we fixed the bypass period length to 2,000 instructions, and then recorded the fraction of the execution time that the system spent in each of the four phases: normal operation, synchronization for checkpoint, recovery and bypass. The results of the study are shown in Figure 8.7, where we also plot the average over the thirteen design bugs and report again the bug occurrence frequencies from Table 8.1 for reference. As the experiment demonstrates, for more frequent bugs, a significant portion of the execution is spent in recovery and bypass, while, for rare errors, this fraction is negligible. Note that all errors were detected by Caspar precisely, since they were all associated with unique transitions of the caches' finite state machines. In this study we observed that, sometimes, several detections were required to bypass a single bug instance. This was the case when the error occurred late in an epoch and a 2,000-instruction bypass was insufficient to circumvent the problem within the first recovery. We also observed situations where several error situations, clustered nearby each other in program execution, were circumvented with just one bypass. Therefore, longer bypass sequences may be beneficial for bugs that are frequent or tend to occur in groups.

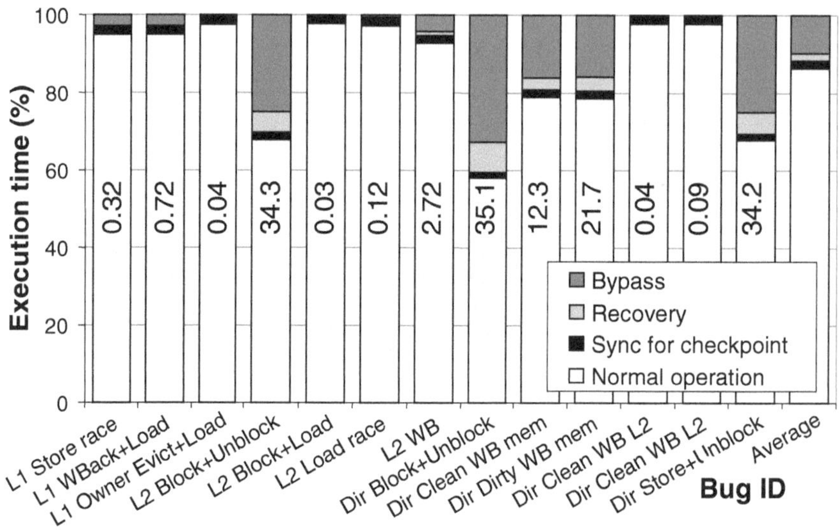

Fig. 8.7 Breakdown of execution time. Fraction of time spent in normal execution, checkpointing, recovery and bypass, for each of the thirteen design bugs over all benchmarks. Note that for errors that have high occurrence rate (as indicated by the frequency values) recovery and bypass constitute a large fraction of the total execution time, while for rare bugs this fraction is negligible.

To analyze the impact of the bypass sequence length on the system's performance we conducted an additional study on five of the design bugs with a widely varied frequency of occurrence. In this experiment the bypass length varied from 500 to 5,000 instructions and the overall performance slowdown of the system was recorded (see Figure 8.8). As it can be noted in the figure, longer bypass intervals are beneficial for frequent errors, primarily because they allow to circumvent an error at the first recovery phase and they have a higher chance of covering multiple errors in a single bypass. On the other hand, we notice an opposite trend for rare bugs (*L1 WBack+Load* and *L2 Block+Load*). In these cases, errors instances are significantly farther from each other, thus a longer bypass period did not reduce the number of recoveries and only incurred higher performance overhead. From this study we conclude that Caspar's fixed-length bypass strategy is not optimal for all bugs. Indeed, the system could be improved to load the number of instructions to run in bypass together with each bug pattern or, to calculate the bypass length dynamically based on where the last error was detected from the beginning of the epoch.

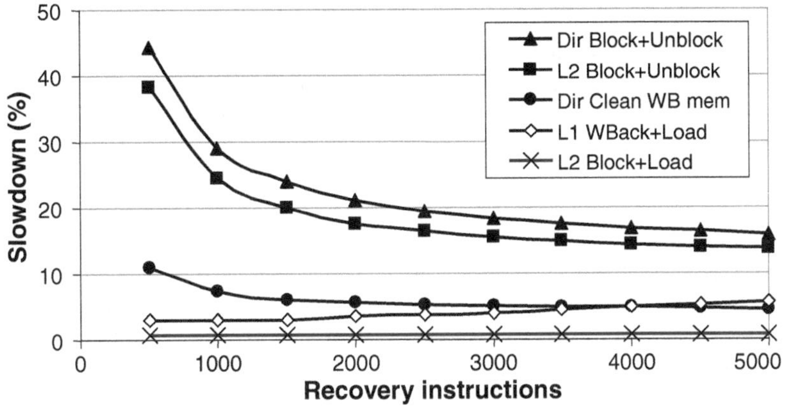

Fig. 8.8 Performance vs. bypass length. For frequently occurring errors, a longer bypass interval is more beneficial, for it allows multiple bug instances to be covered together. For rare bugs, on the other hand, a longer bypass results in larger penalties due to longer execution at lower performance.

8.5.4 Area Overhead

Finally, we investigated the area overhead of Caspar's event detectors. To this end, we implemented the detectors for the L1 and L2 caches in Verilog HDL and synthesized them using the TSMC 90nm technology. The area of the L1 and L2 cache detectors was found to be $984.2\mu m^2$ and $2{,}422\mu m^2$, respectively. In comparison, an OpenSPARC T1 design, which has eight cores with private caches and a shared four-bank L2 cache, has an approximate area of $378mm^2$ [LTS$^+$07]. Thus, placing event detectors in each private cache and in each L2 bank incurs an area overhead of 0.005%, a negligible penalty compared to the area of the checkpointing log storage analyzed in Section 8.5.2. Considering that the area overhead of the checkpointing is calculated as a percentage of the area of the caches, we find that in the OpenSPARC T1 design the checkpointing logs and detectors of Caspar will occupy less than 3% of the total processor die area.

8.6 Summary

In this chapter we overviewed two runtime verification solutions designed specifically for modern multi-core processors. The first technique, Dynamic Verification of Memory Consistency (DVMC), is based on non-programmable checkers that enforce three invariants common to all memory consistency models: *uniprocessor ordering, allowed reordering* and *cache coherence*. When a violation to an invariant is detected, DVMC uses a checkpointing technique to roll back the processor's state and restart execution. The second solution presented, called Caspar, was designed by us for efficient in-the-field patching and repair of the memory subsystem in multi-core processors. As DVMC, Caspar relies on periodic system checkpointing, but uses programmable event detectors instead of hardware checkers to identify erroneous transitions in the cache controllers' state machines. When an error is detected during the processor's operation, Caspar suspends the execution, recovers the last correct checkpointed system state, and attempts to bypass the bug by enforcing race-free memory activity. Our experimental results demonstrate that Caspar's checkpointing scheme incurs little performance and area penalty (2.4% and 5.8%, respectively, for the target system in our experiments). The DVMC technique, on the other hand, relies on a more heavyweight checkpointing scheme (SafetyNet), which has lower performance impact, but occupies a larger fraction of a silicon die. Nevertheless, with no errors present in the design, the performance impact of Caspar (due to checkpointing) is still quite small. When errors do occur, recovery can potentially become expensive; however, we found experimentally that with low error frequencies (which is expected in commercial systems that undergo heavy pre- and post-silicon verification) the overall performance penalty is also small. These features allow Caspar to work as an effective patching solution, protecting even the most complex of today's designs.

References

[LTS+07] Ana S. Leon, Kenway W. Tam, Jinuk L. Shin, David Weisner, and Francis Schumacher. A power-efficient high-throughput 32-thread SPARC processor. *IEEE Journal of Solid-State Circuits*, 42(1):7–16, January 2007.

[MCE+02] Peter S. Magnusson, Magnus Christensson, Jesper Eskilson, Daniel Forsgren, Gustav Hallberg, Johan Hogberg, Frederik Larsson, Andreas Moestedt, and Bengt Werner. Simics: A full system simulation platform. *IEEE Computer*, 35(2):50–58, 2002.

[MS05] Albert Meixner and Daniel J. Sorin. Dynamic verification of sequential consistency. *SIGARCH Computer Architecture News*, 33(2):482–493, 2005.

[MS06] Albert Meixner and Daniel J. Sorin. Dynamic verification of memory consistency in cache-coherent multithreaded computer architectures. In *DSN, Proceedings of the International Conference on Dependable Systems and Networks*, pages 73–82, June 2006.

[MSB+05] Milo M. K. Martin, Daniel J. Sorin, Bradford M. Beckmann, Michael R. Marty, Min Xu, Alaa R. Alameldeen, Kevin E. Moore, Mark D. Hill, and David A. Wood. Multifacet's general execution-driven multiprocessor simulator (GEMS) toolset. *SIGARCH Computer Architecture News*, 33(4):92–99, 2005.

[NMGT06] Jun Nakano, Pablo Montesinos, Kourosh Gharachorloo, and Josep Torrellas. ReViveI/O: efficient handling of I/O in highly-available rollback-recovery servers. In *International Symposium on High-Performance Computer Architecture*, pages 203–214, February 2006.

[SMHW02] Daniel J. Sorin, Milo Martin, Mark Hill, and David Wood. SafetyNet: improving the availability of shared memory multiprocessors with global checkpoint/recovery. In *ISCA, Proceedings of the International Symposium on Computer Architecture*, pages 123–134, 2002.

[WOT+95] Steven Woo, Moriyoshi Ohara, Evan Torrie, Jaswinder Singh, and Anoop Gupta. The SPLASH-2 programs: characterization and methodological considerations. In *ISCA, Proceedings of the International Symposium on Computer Architecture*, pages 24–36, June 1995.

Chapter 9
ENSURING CORRECTNESS IN FUTURE MICROPROCESSORS

Abstract. In the concluding chapter of this book we present our outlook at the future of processor verification. As the complexity of microprocessors grows and volumes of production increase, hardware vendors are subjected to greater risks of showstopper bugs. Yet, when it comes to verification, the microprocessor industry is often reactive, developing new solutions after critical escapes occur. Here, we advocate a proactive approach to verification, where the chip's architecture is analyzed early in the design phase and pre-silicon, post-silicon and runtime techniques are integrated and developed according to a design's needs. In this chapter we also overview upcoming trends in microprocessor architectures, such as heterogeneous multi-core structures and complex on-chip interconnects, and discuss how these trends will affect the nature of processor validation.

9.1 Advances and Trends in Processor Validation

Over the last four decades, advances in processor validation followed the evolution of the designs under verification themselves, rapidly increasing in capabilities and scope. Yet, as we discussed throughout this book, our ability to design digital circuits is constantly outpacing our ability to verify them, widening the verification gap with every new generation of microprocessors. The response of chip designing companies to this trend has been to employ mostly extensive, rather than intensive approaches. In other words, verification teams and capital investments have grown, while methods of validation have shown little change. Quantum improvements in verification technology and new techniques have been developed most often in response to the discovery of critical escape errors that had a major impact on a processor product. For example, the FDIV bug that escaped in the Intel Pentium processor, and the subsequent recall of defective parts have led to significant profit losses and caused a wave of negative publicity for Intel. This highlighted to all hardware ven-

I. Wagner and V. Bertacco, *Post-Silicon and Runtime Verification for Modern Processors*,
DOI 10.1007/978-1-4419-8034-2_9, © Springer Science+Business Media, LLC 2011

dors the value of employing formal pre-silicon verification practices, which could have caught this error, in their design flows. As a response, Intel staffed its teams with several additional verification engineers that focused exclusively on formal verification aspects, particularly of datapath units, leading to effective contributions (in terms of discovering additional bugs early in the development phase) in subsequent product generations.

Similarly, escaped control logic errors showed that without microcode patching techniques, the only viable options left to manufacturers upon discovery of a bug were to recall the faulty components or to completely disable the affected on-die features, often leading to severe performance loss. The 2007 escaped bug in AMD's newest Phenom processor, which was related to the memory subsystem, is undoubtedly pushing the industry to step up verification efforts in cache-hierarchy and on-chip interconnect, as well as to consider solutions that can ensure the runtime correctness of memory operations. This bug and the trends towards increasing the number of cores in a single chip, which in turn leads to extremely complex memory subsystems, has already stimulated much research in academic environments and has advocated the development of several of the techniques presented in this book.

9.2 A Proactive Approach to Verification

We believe, however, that in general a *reactive* approach to verification, where actions are taken only after a major slip-up, is fundamentally flawed: the financial blow of a nasty escape to a manufacturing company may be too much to recover with subsequent products. Production volumes today are much higher than, for instance, in the days of the first Pentium generation, thus recalling millions of defected microprocessors would cost a company today significantly more than the $420 millions reported by Intel in 1994. Delays in the release of a new processor have similar effects in today's competitive markets. Therefore, hardware validation must be *proactive*: researchers in industry must analyze new designs for error vulnerabilities, anticipate the types of errors that may occur in them and design verification techniques to fulfill the project needs. To some degree, companies like Intel have already started to deploy this early validation analysis, however, more proactive steps must be taken for future generations, when designs become even more complex.

In addition to complexity and device density growth, recently developed processors are also incorporating new features, never tackled by verification before. As with all cutting-edge, untested technologies, these features present greater exposure to escaped bugs and, therefore, require the development of novel validation techniques, geared specifically for them. For example, in this book we had several discussions on validation solutions specific to multi-core designs. Before the appearance of this processor architecture in the early 2000s, much of vendors' verification resources were spent on a device's computational engine. In multi-cores, however, the focus has shifted, because datapaths are simpler and thus are easier to validate, while the communication between cores through the memory subsystem has notably increased in complexity. Therefore, novel approaches for pre-silicon, post-silicon and runtime verification of cache coherence and memory consistency have been and are being developed. In the future, with the number of cores in a chip expected to increase to tens and even hundreds, the importance and impact of the interconnect subsystem will also grow, and, consequently, its validation will require a much more significant share of a company's resources.

It is also expected that complex interconnects in multi-core designs will also require scope of verification to be extended to include statistical validation of *quality of service* (QoS). With more and more diverse components being integrated onto a processor die, today's microprocessors begin to look more like systems on a chip (SoC), than the traditional computing devices of the past. For instance, the latest CPU generations integrate on-chip on-die memory controllers and dedicated modules to communicate serially with peripheral devices. Furthermore, it is projected that individual processor cores will become more application-specific in the future. The initial stages of this trend can already be noted today with the integration of graphics processing units and cryptographic cores on the same silicon die as the general purpose processor. Consequently, on-chip communication protocols will have to adapt to allow for different classes of traffic, each necessitating specific guarantees for quality of service. For instance, data exchanges between general purpose computational cores should be protected by guarantees of reliable delivery and maximum acceptable latency to ensure the absence of deadlocks and inconsistent memory states. In contrast, traffic streams to and from a graphics core or a hardware video codec (COmpressor-DECompressor) can be delayed, and it is often acceptable to even drop data packets in case of contention. As a result, a video output could become jittery, degrading user experience; however, the overall correctness of the system's computation would not be compromised. A similar reasoning can be made for the reliability of the data delivered to the various units: indeed, video streams can typically endure a much higher error rate than control and general purpose computation traffic. To address such situations, engineers could overdesign the on-die network to effectively cope with peak amounts of traffic, eliminating the possibility of contention in the network altogether. The price of this choice, however, may be a dramatic decrease in the power efficiency of the device, even to the point of being unviable. An alternative approach is to provide QoS capabilities in delivering different message streams, similar to what is done today in the networking domain, where different classes of traffic are provided statistically with different latency guarantees. For instance, a typical quality of service may enforce a router constraint to forward no more than 60% of low priority traffic when the total influx into the router exceeds a certain threshold. Such a constraint allows costly on-die networking resources to be used much more efficiently. However, engineers validating this type of system will then be faced with the prospect of verifying these statistical guarantees for individual routers, as well as for the entire network-on-chip. Formal verification, in this case, may benefit greatly from advances incorporating queueing theory, while self-guiding test generators will need to obtain feedback from on-chip routing nodes to be able to stress various QoS operation modes in pre-silicon models and early hardware prototypes.

Runtime verification is also bound to become an important factor in future designs, as semiconductor technology continues to shrink the size of on-die features. As transistors become smaller, they also become less reliable and more susceptible to dynamic errors, *e.g.*, energetic particle strikes, voltage noise, *etc.* Moreover, with smaller feature sizes, the average lifetime of individual transistors decreases, and circuits made of billions of such unreliable components become likely to fail within just a few years. It is important to note here, that if no protection measures are taken, the failure of a single transistor can cause the entire system to malfunction and become unusable. Today, silicon wearout issues are addressed with a host of techniques in both the design and manufacturing domains. In the latter, freshly fabricated silicon dies are subjected to burn-in tests, in which devices are overclocked and overheated while they are being subjected to stressful computational stimuli: the goal is to make the weakest transistors fail rapidly. Engineers then validate that a device is still functioning correctly after the burn-in test. When that is the case, some assurance can be attained that the silicon part will not break in a short term.

In addition, to ensure that individual transistor faults do not render an entire processor inoperable, designers often add redundant modules to the system that can be enabled upon failure of main processing blocks. Redundancy can, therefore, provide a reliable recovery mechanism, however, it becomes crucial in this case to detect and diagnose errors in a timely manner at runtime. This is where runtime validation techniques, which check the correctness of a device's operation in the field, can play a vital role. These solutions can silently monitor the health of a computer system and invoke maintenance, reconfiguration and repair algorithms, if the effects of aging begin to appear and manifest themselves as functional errors. Similarly, runtime techniques can provide high-quality protection from dynamic errors (led by transient errors in transistors), which may corrupt computation results and/or communication messages. In fact, the majority of the runtime techniques presented in Chapter 6 include dynamic error detection and correction as one of their primary goals. Pre- and post-silicon verification approaches, on the other hand, need to be extended to validate the performance of dynamic solutions at design time, so error detection and recovery hardware itself is reliable and correct. Overall, with semiconductor technology moving deeper into the sub-micron domain, we expect that verification and reliability concerns will be merged together, and errors of various origins, such as functional, transient and wearout-related, will all be handled efficiently by the same robust mechanisms.

Finally, future processors will probably include components from outside the traditional CMOS domain, which will also need to be verified to interact correctly with the CPU's digital logic. Micro-electro-mechanical devices and biological labs-on-a-chip are, perhaps, two of the most intriguing examples of technologies that may be incorporated into future processors. Verification of the internal functionality of these exotic components and their communication with the rest of the processor logic will require engineering knowledge from mechanics, physics, chemistry and biology. Moreover, simulation and formal analysis tools will have to be augmented to take these novel technologies into account. Likewise, post-silicon validation for such components will be extended into domains of mechanical and biological testing. These examples are, however, somewhat extreme and it is more likely that the majority of processors will avoid such unusual peripherals. Nevertheless, as we pointed out before, on-chip components in modern processors are becoming heterogenous, and to cope with this growing diversification validation must be fused with techniques from other fields. Most importantly, however, engineers in industry and academia alike must always beware of growing design complexity and look into novel solutions to try to close the verification gap.

References

[AAF+04] Allon Adir, Eli Almog, Laurent Fournier, Eitan Marcus, Michal Rimon, Michael Vinov, and Avi Ziv. Genesys-pro: Innovations in test program generation for functional processor verification. *IEEE Design & Test of Computers*, 21(2):84–93, 2004.

[ABD+06] Miron Abramovici, Paul Bradley, Kumar Dwarakanath, Peter Levin, Gerard Memmi, and Dave Miller. A reconfigurable design-for-debug infrastructure for SoCs. In *DAC, Proceedings of the Design Automation Conference*, pages 7–12, July 2006.

[ABF94] Miron Abramovici, Melvin A. Breuer, and Arthur D. Friedman. *Digital Systems Testing and Testable Design*. Wiley-IEEE Press, revised edition, September 1994.

[Acc04] Accellera. *Property Specification Language Reference Manual, Rev 1.1*, June 2004. `http://www.eda.org/vfv/docs/PSL-v1.1.pdf`.

[AMD03] Advanced Micro Devices, Inc. *AMD Athlon™ Processor Model 6 Revision Guide*, October 2003. `http://www.amd.com/us-en/assets/content_type/white_papers_and_tech_docs/24332.pdf`.

[AMD05] Advanced Micro Devices, Inc. *Revision Guide for AMD Athlon™ 64 and AMD Opteron™ Processors*, August 2005. `http://www.amd.com/us-en/assets/content_type/white_papers_and_tech_docs/25759.pdf`.

[AMD08] Advanced Micro Devices, Inc. *Revision Guide for AMD Family 10h Processors*, April 2008. `http://.amd.com/us-en/assets/content_type/white_papers_and_tech_docs/41322.PDF`.

[AMD09] AMD Phenom™ X4 Quad-Core and AMD Phenom™ X3 Triple-Core Processors for Home, 2009. `http://www.amd.com/us-en/Processors/ProductInformation/0,,30_118_15331_15332,00.html`.

[AMS08] Charles J. Alpert, Dinesh P. Mehta, and Sachin S. Sapatnekar. *Handbook of Algorithms for Physical Automation*. Taylor and Francis, first edition, 2008.

[Arv03] Arvind. Bluespec: A language for hardware design, simulation, synthesis and verification. In *MEMOCODE, Proceedings of the ACM/EEE International Conference on Formal Methods and Models for Co-Design*, page 249, June 2003.

[ASL03] Dennis Abts, Steve Scott, and David J. Lilja. So many states, so little time: Verifying memory coherence in the Cray X1. In *IPDPS, Proceedings of the International Symposium on Parallel and Distributed Processing*, page 11.2, April 2003.

[Aus00] Todd Austin. DIVA: A dynamic approach to microprocessor verification. *Journal of Instruction-Level Parallelism*, 2:1–26, May 2000.

[BA00] Michael Bushnell and Vishwanti Agrawal. *Essentials of Electronic Testing for Digital, Memory and Mixed-signal VLSI Circuits*. Springer, 2000.

[BA08] David Burger and Todd Austin. The SimpleScalar toolset, version 3.0, 2008. `http://simplescalar.com`.

I. Wagner and V. Bertacco, *Post-Silicon and Runtime Verification for Modern Processors*,
DOI 10.1007/978-1-4419-8034-2, © Springer Science+Business Media, LLC 2011

[BASS07] Tommy Bojan, Manuel A. Arreola, Eran Shlomo, and Tal Shachar. Functional coverage measurements and results in post-silicon validation of Core™2 Duo family. In *HLDVT, Proceedings of the International Workshop on High Level Design Validation and Test*, pages 145–150, November 2007.

[BAWR07] Jayanta Bhadra, Magdy S. Abadir, Li-C. Wang, and Sandip Ray. A survey of hybrid techniques for functional verification. *IEEE Design and Test of Computers*, 24(2):112–122, March 2007.

[BBB+87] Randal E. Bryant, Derek L. Beatty, Karl Brace, Kyeongsoon Cho, and Thomas Sheffler. COSMOS: A compiled simulator for MOS circuits. In *DAC, Proceedings of the Design Automation Conference*, pages 9–16, June 1987.

[BBS90] Derek L. Beatty, Randal E. Bryant, and Carl-Johann H. Seger. Synchronous circuit verification by symbolic simulation: An illustration. In *Proceedings of Sixth MIT conference on advanced research in VLSI*, pages 98–112, April 1990.

[BDF+03] Paul Barham, Boris Dragovic, Keir Fraser, Steven Hand, Tim Harris, Alex Ho, Rolf Neugebauer, Ian Pratt, and Andrew Warfield. Xen and the art of virtualization. In *SOSP, Proceedings of the ACM Symposium on Operating Systems Principles*, pages 164–177, October 2003.

[BEA+08] Shane Bell, Bruce Edwards, John Amann, Rich Conlin, Kevin Joyce, Vince Leung, John MacKay, Mike Reif, Liewei Bao, John Brown, Matthew Mattina, Chyi-Chang Miao, Carl Ramey, David Wentzlaff, Walker Anderson, Ethan Berger, Nat Fairbanks, Durlov Khan, Froilan Montenegro, Jay Stickney, and John Zook. TILE64™processor: A 64-core SoC with mesh interconnect. In *ISSCC, Proceedings of the International Solid State Circuits Conference*, pages 88–598, February 2008.

[BEC06] Keith H. Bierman, David R. Emberson, and Liang T. Chen. *U.S. Patent no. 7133818: Method and apparatus for accelerated post-silicon testing and random number generation.* Sun Microsystems, Inc., November 2006.

[Ben01] Bob Bentley. Validating the Intel® Pentium® 4 microprocessor. In *DAC, Proceedings of the Design Automation Conference*, pages 224–228, June 2001.

[Ben05] Bob Bentley. Validating a modern microprocessor. In *CAV, Proceedings of the International Conference on Computer Aided Verification*, pages 2–4, July 2005.

[Ber05] Valeria Bertacco. *Formal Verification Scalable Hardware Verification With Symbolic Simulation.* Springer, first edition, 2005.

[BHMSV84] Robert K. Brayton, Gary D. Hachtel, Curtis T. McMullen, and Alberto L. Sangiovanni-Vincentelli. *Logic Minimization Algorithms for VLSI Synthesis.* Kluwer Academic Publishers, first edition, 1984.

[BHS99] Eric B. Brett, David P. Hunter, and Sharon L. Smith. Moving Atom to Windows NT for Alpha. *Compaq DIGITAL Technical Journal*, 10(2):1–26, January 1999.

[BLL+04] Michael Behm, John Ludden, Yossi Lichtenstein, Michal Rimon, and Michael Vinov. Industrial experience with test generation languages for processor verification. *DAC, Proceedings of the Design Automation Conference*, pages 36–40, June 2004.

[BM05] Ali A. Bayazit and Sharad Malik. Complementary use of runtime validation and model checking. In *ICCAD, Proceedings of the International Conference on Computer Aided Design*, pages 1052–1059, 2005.

[Boc07] Bochs: The open source IA-32 emulation project, September 2007. `http://bochs.sourceforge.net/`.

[BP80] Henry Baker and Clinton Parker. High level language programs run ten times faster in microstore. *ACM SIGMICRO Newsletter*, 11(3-4):171–177, 1980.

[BRB90] Karl Brace, Richard Rudell, and Randal E. Bryant. Efficient implementation of a BDD package. In *DAC, Proceedings of the Design Automation Conference*, pages 40–45, June 1990.

[Bry85] Randal E. Bryant. Symbolic verification of MOS circuits. In *Proceedings of Chapel Hill Conference on VLSI*, pages 419–438, May 1985.

[Bry86] Randal E. Bryant. Graph-based algorithms for Boolean function manipulation. *IEEE Transactions on Computers*, 35:677–691, 1986.

[Car90] Adrian Carbine. *U.S. Patent no. 5253255: Scan mechanism for monitoring the state of internal signals of a VLSI microprocessor chip*. Intel Corporation, November 1990.

[CBM89] Oliver Coudert, Christian Berthet, and Jean C. Madre. Verification of synchronous sequential machines based on symbolic execution. In *Proceedings of the International Workshop on Automatic Verification Methods for Finite State Systems*, pages 365–373, June 1989.

[CBM05] Kai-hui Chang, Valeria Bertacco, and Igor Markov. Simulation-based bug trace minimization with BMC-based refinement. In *ICCAD, Proceedings of the International Conference on Computer Aided Design*, pages 1045–1051, November 2005.

[CBM07a] Kai-hui Chang, Valeria Bertacco, and Igor Markov. Automating post-silicon debugging and repair. In *ICCAD, Proceedings of the International Conference on Computer Aided Design*, pages 91–98, November 2007.

[CBM07b] Kai-hui Chang, Valeria Bertacco, and Igor Markov. Simulation-based bug trace minimization with BMC-based refinement. *IEEE Transactions on Computer-Aided Design of Integrated Circuits and Systems*, 26(1):152–165, 2007.

[CBRZ01] Edmund M. Clarke, Armin Biere, Richard Raimi, and Yunshan Zhu. Bounded model checking using satisfiability solving. *Formal Methods in System Design*, 19(1):7–34, 2001.

[CGP08] Edmund M. Clarke, Orna Grumberg, and Doron A. Peled. *Model Checking*. MIT Press, sixth edition, 2008.

[CJB79] William C. Carter, William H. Joyner, and Daniel Brand. Symbolic simulation for correct machine design. In *DAC, Proceedings of the Design Automation Conference*, pages 280–286, June 1979.

[CKM+04] William D. Corti, Robert Kenny, Joseph O. Marsh, Steven C. Parker, Frank X. Scanzano, and Michael Won. *U.S. Patent no. 6834360: On-chip logic analyzer*. International Business Machines Corporation, December 2004.

[CMA08] Kypros Constantinides, Onur Mutlu, and Todd Austin. Online design bug detection: RTL analysis, flexible mechanisms, and evaluation. In *MICRO, Proceedings of the International Symposium on Microarchitecture*, pages 282–293, November 2008.

[CMH00] David Van Campenhout, Trevor Mudge, and John P. Hayes. Collection and analysis of microprocessor design errors. *IEEE Design and Test of Computers*, 17(4):51–60, 2000.

[CMP08] Kaiyu Chen, Sharad Malik, and Priyadarsan Patra. Runtime validation of memory ordering using constraint graph checking. In *HPCA, Proceedings of the International Symposium on High-Performance Computer Architecture*, pages 415–426, February 2008.

[CSG98] David Culler, Jaswinder P. Singh, and Anoop Gupta. *Parallel Computer Architecture: A Hardware/Software Approach*. Morgan Kaufmann, first edition, 1998.

[CVK04] Ben Cohen, Srinivasan Venkataramanan, and Ajeetha Kumari. *Using PSL/Sugar for Formal and Dynamic Verification*. VhdlCohen Publishing, second edition, 2004.

[CWBM07] Kai-hui Chang, Ilya Wagner, Valeria Bertacco, and Igor Markov. Automatic error diagnosis and correction for RTL designs. In *HLDVT, Proceedings of the International Workshop on High Level Design Validation and Test*, pages 65–72, November 2007.

[CYGC06] Xiaofang Chen, Yu Yang, Ganesh Gopalakrishnan, and Ching-Tsun Chou. Reducing verification complexity of a multicore coherence protocol using assume/guarantee. In *FMCAD, Proceedings of the Formal Methods in Computer Aided Design Conference*, pages 81–88, November 2006.

[DBB07] Andrew DeOrio, Adam Bauserman, and Valeria Bertacco. Chico: An on-chip hardware checker for pipeline control logic. In *MTV, Proceedings of the International Workshop on Microprocessor Test and Verification*, pages 91–97, December 2007.

[DBB08] Andrew DeOrio, Adam Bauserman, and Valeria Bertacco. Post-silicon verification for cache coherence. In *ICCD, Proceedings of the International Conference on Computer Design*, pages 348–355, October 2008.

[DDHY92] David L. Dill, Andreas J. Drexler, Alan J. Hu, and C. Han Yang. Protocol verification as a hardware design aid. In *ICCD, Proceedings of the International Conference on Computer Design*, pages 522–525, October 1992.

[DDJ06] DDJ Microprocessor Center, 2006. `http://www.x86.org/`.

[DP60] Martin Davis and Hilary Putnam. A computing procedure for quantification theory. *Journal of the Association for Computing Machinery*, 7(3):201–215, July 1960.

[DWB09] Andrew DeOrio, Ilya Wagner, and Valeria Bertacco. Dacota: Post-silicon validation of the memory subsystem in multi-core designs. In *HPCA, Proceedings of the International Symposium on High-Performance Computer Architecture*, pages 405–416, February 2009.

[ES03] Niklas Een and Niklas Sorensson. An extensible SAT-solver. In *Theory and Applications of Satisfiability Testing*, volume 2919 of *Lecture Notes in Computer Science*, pages 502–518. Springer, 2003.

[EWR00] Travis Eiles, Gary Woods, and Valluri Rao. Optical probing of flip-chip-packaged microprocessors. In *ISSCC, Proceedings of the International Solid State Circuits Conference*, pages 220–221, February 2000.

[FKL04] Harry Foster, Adam Krolnik, and David Lacey. *Assertion-based design.* Springer, second edition, 2004.

[FZ03] Shai Fine and Avi Ziv. Coverage directed test generation for functional verification using Bayesian networks. In *DAC, Proceedings of the Design Automation Conference*, pages 286–281, June 2003.

[GC95] Michael D. Goddard and David S. Christie. *U.S. Patent no. 5796974: Microcode patching apparatus and method.* Advanced Micro Devices, Inc., November 1995.

[Ger03] Steven German. Formal design of cache memory protocols in IBM. *Formal Methods in System Design*, 22(2):133–141, 2003.

[Glu06] Alon Gluska. Practical methods in coverage-oriented verification of the Merom microprocessor. In *DAC, Proceedings of the Design Automation Conference*, pages 332–337, July 2006.

[GTKS05] Mehmet H. Gunes, Mitchell Thornton, Fatih Kocan, and Stephen Szygenda. A and comparison of digital logic simulators. In *MWSCAS, Proceedings of the Midwest Symposium on Circuits and Systems*, pages 744–749, August 2005.

[HEL$^+$86] Mark D. Hill, Susan Eggers, Jim Larus, George Taylor, Glenn Adams, B. K. Bose, Garth A. Gibson, Paul Hansen, Jon Keller, Shing Kong, Corinna Lee, Daebum Lee, Joan Pendleton, Scott Ritchie, David A. Wood, Ben Zorn, Paul Hilfinger, Dave Hodges, Randy H. Katz, John Ousterhout, and Dave Patterson. Design decisions in SPUR. *IEEE Computer*, 19(11):8–22, November 1986.

[HKM01] Faisal I. Haque, Khizar A. Khan, and Jonathan Michelson. *The Art of Verification with Vera.* Verification Central, first edition, 2001.

[HMN01] Yoav Hollander, Matthew Morley, and Amos Noy. The *e* language: A fresh separation of concerns. In *TOOLS, Proceedings of the, International Conference on Technology of Object-Oriented Languages*, pages 41–50, March 2001.

[HS06] Gary D. Hachtel and Fabio Somenzi. *Logic Synthesis and Verification Algorithms.* Springer, first edition, 2006.

[HSH$^+$00] Pei-Hsin Ho, Thomas Shiple, Kevin Harer, James Kukula, Robert Damiano, Valeria Bertacco, Jerry Taylor, and Jiang Long. Smart simulation using collaborative formal and simulation engines. In *ICCAD, Proceedings of the International Conference on Computer Aided Design*, pages 120–126, November 2000.

[HTB$^+$09] Richard Ho, Michael Theobald, Brannon Batson, J.P. Grossman, Stanley C. Wang, Joseph Gagliardo, Martin M. Deneroff, Ron O. Dror, and David E. Shaw. Post-silicon debug using formal verification waypoints. In *DVCon, Proceedings of the Design and Verification Conference and Exhibition*, pages 1–7, February 2009.

[IBM05] International Business Machines Corporation. *IBM PowerPC 750GX and 750GL RISC Microprocessor Errata Notice*, July 2005. `http://www-01.ibm.com/chips/techlib/techlib.nsf/techdocs/CC0918EAD88CB09E87256F5C006FF9E6`.

[IEE01a] Institute of Electrical and Electronics Engineers. *Standard test access port and boundary-scan architecture. IEEE Std. 1149.1-2001.*, 2001. http://ieeexplore.ieee.org/servlet/opac?punumber=7481.

[IEE01b] Institute of Electrical and Electronics Engineers. *Standard Verilog hardware description language. Std 1364-2001.*, 2001. http://ieeexplore.ieee.org/servlet/opac?punumber=7578.

[IEE04] Institute of Electrical and Electronics Engineers. *Behavioural languages - Part 1-1: VHDL language reference manual. IEC 61691-1-1 First edition 2004-10; IEEE 1076.*, 2004. http://ieeexplore.ieee.org/servlet/opac?punumber=9649.

[IEE07] Institute of Electrical and Electronics Engineers. *Standard for SystemVerilog - Unified Hardware Design, Specification, and Verification Language. Std 1800-2007*, 2007. http://ieeexplore.ieee.org/servlet/opac?punumber=4410438.

[Int00] Intel Corporation. *Intel®StrongARM®SA-1100 Microprocessor Specification Update*, February 2000.

[Int02a] Intel Corporation. *Intel®Celeron®Processor Specification Update*, September 2002. http://developer.intel.com/design/celeron/specupdt/24374847.pdf.

[Int02b] Intel Corporation. *Intel®Pentium®II Processor Invalid Instruction Erratum Overview*, July 2002. http://developer.intel.com/design/pentiumii/specupdt/24333749.pdf.

[Int04] Intel Corporation. *Intel®Pentium®Processor Invalid Instruction Erratum Overview*, July 2004. http://www.intel.com/support/processors/pentium/sb/cs-013151.htm.

[Int05] Intel Corporation. *Intel®Pentium®III Processor Specification Update*, May 2005. http://download.intel.com/design/PentiumIII/specupdt/24445355.pdf.

[Int07a] Intel Corporation. *Intel®Core™Duo Desktop Processor E6000 and E4000 Sequence Specification Update*, November 2007. http://download.intel.com/design/processor/specupdt/31327921.pdf.

[Int07b] Intel Corporation. *Intel®Core™Extreme Quad-Core Processor QX6000 Sequence and Intel®Core™Quad Processor Q6000 Sequence*, November 2007. http://download.intel.com/design/processor/specupdt/31559318.pdf.

[Int08] Intel Corporation. *Intel®Core™2 Duo and Intel®Core™2 Solo Processor for Intel®Centrino®Duo Processor Technology Specification Update*, October 2008. http://download.intel.com/design/mobile/SPECUPDT/31407918.pdf.

[Int09] Intel Corporation. *Intel®Core™Solo processor - Intel®Core™Solo Processor support*, 2009. http://www.intel.com/support/processors/mobile/coresolo/.

[ITR07] International Technology Roadmap for Semiconductors executive summary, 2007. http://www.itrs.net/Links/2007ITRS/Home2007.htm.

[JG04] Doug Josephson and Bob Gottlieb. The crazy mixed up world of silicon debug. In *CICC, Proceedings of the IEEE Custom Integrated Circuits Conference*, pages 665–670, October 2004.

[JPG01] Doug Josephson, Steve Poehhnan, and Vincent Govan. Debug methodology for the McKinley processor. In *ITC, Proceedings of the International Test Conference*, pages 451–460, October 2001.

[KAI07] Jai Kumar, Catherine Ahlschlager, and Peter Isberg. Post-silicon verification methodology on Sun's UltraSPARC T2 processor. In *HLDVT, Proceedings of the International Workshop on High Level Design Validation and Test*, page 47, November 2007.

[KAO05] Poonacha Kongetira, Kathirgamar Aingaran, and Kunle Olukotun. Niagara: a 32-way multithreaded SPARC processor. *IEEE Micro*, 25(2):21 – 29, March 2005.

[Kas08] Kris Kaspersky. Remote code execution through Intel CPU bugs. In *HITB, Proceedings of Hack In The Box Conference*, October 2008.

[KDF+04] Ravishankar Kuppuswamy, Peter DesRosier, Derek Feltham, Rehan Sheikh, and Paul Thadikaran. Full hold-scan systems in microprocessors: Cost/benefit analysis. *Intel Technology Journal*, 08:63–72, February 2004.

[KDH+05] James A. Kahle, Michael N. Day, H. Peter Hofstee, Charles R. Johns, Thodore R. Maeurer, and David Shippy. Introduction to the Cell multiprocessor. *IBM Journal of Research and Development*, 49:589–604, 2005.

[KG99] Christoph Kern and Mark R. Greenstreet. Formal verification in hardware design: a survey. *ACM Transactions on Design Automation of Electronic Systems*, 4(2):123–193, 1999.

[KK94] Ramayya Kumar and Thomas Kropf. *Theorem provers in circuit design: theory, practice, and experience*. Springer Berlin/ Heidelberg, first edition, 1994.

[Knu97] Donald E. Knuth. *The Art of Computer Programming*. Addison-Wesley Professional, third edition, 1997.

[Kon09] Dusko Koncaliev. Bugs in the Intel microprocessors, 2009. `http://www.cs.earlham.edu/~dusko/cs63/`.

[Lam97] Leslie Lamport. How to make a correct multiprocess program execute correctly on a multiprocessor. *IEEE Transactions on Computers*, 46(7):779–782, 1997.

[Lam02] Leslie Lamport. *Specifying systems: the TLA+ language and tools for hardware and software engineers*. Addison-Wesley Professional, 2002.

[Lit02] Timothe Litt. Support for debugging in the Alpha 21364 microprocessor. In *ITC, Proceedings of the International Test Conference*, pages 584–589, October 2002.

[LTS+07] Ana S. Leon, Kenway W. Tam, Jinuk L. Shin, David Weisner, and Francis Schumacher. A power-efficient high-throughput 32-thread SPARC processor. *IEEE Journal of Solid-State Circuits*, 42(1):7–16, January 2007.

[M507] The M5 simulator system, November 2007. `http://www.m5sim.org`.

[Mar08] John Markoff. Burned once, Intel prepares new chip fortified by constant tests. *New York Times*, November 2008. `http://www.nytimes.com/2008/11/17/technology/companies/17chip.html`.

[MBS08] Albert Meixner, Michael E. Bauer, and Daniel J. Sorin. Argus: Low-cost, comprehensive error detection in simple cores. *IEEE Micro*, 28(1):52–59, 2008.

[McC56] Edward J. McCluskey. Minimization of Boolean functions. *Bell Systems Technical Journal*, 6(35):1417–1444, November 1956.

[MCE+02] Peter S. Magnusson, Magnus Christensson, Jesper Eskilson, Daniel Forsgren, Gustav Hallberg, Johan Hogberg, Frederik Larsson, Andreas Moestedt, and Bengt Werner. Simics: A full system simulation platform. *IEEE Computer*, 35(2):50–58, 2002.

[McG07] Harlan McGhan. The gHost in the machine. *Microprocessor Report*, 21(3):11–33, March 2007.

[MDM+06] Massimiliano Melani, Francesco D'Ascoli, Corrado Marino, Luca Fanucci, Adolfo Giambastiani, Alessandro Rochhi, Marco De Marinis, and Andrea Monterastelli. An integrated flow from pre-silicon simulation to post-silicon verification. In *Research in Microelectronics and Electronics 2006, Ph.D.*, pages 205–208, June 2006.

[Mei93] Gerd Meister. A survey on parallel logic simulation. Technical report, University of Saarland, Department of Computer Science, 1993.

[Men08] Mentor Graphics®Inc. *ModelSim - a comprehensive simulation and debug environment for complex ASIC and FPGA designs*, 2008. `http://www.model.com/`.

[Mic03] Alexander Miczo. *Digital logic testing and simulation*. John Wiley and Sons, second edition, 2003.

[Mic09] Microsoft Corporation. */QIfdiv (Enable Pentium®FDIV Fix)*, 2009. `http://msdn2.microsoft.com/en-us/library/ms856573.aspx`.

[MP99] Kevin J. McGrath and James K. Pickett. *U.S. Patent no. 6438664: Microcode patch device and method for patching microcode using match registers and patch routines*. Advanced Micro Devices, Inc., October 1999.

[MS05] Albert Meixner and Daniel J. Sorin. Dynamic verification of sequential consistency. *SIGARCH Computer Architecture News*, 33(2):482–493, 2005.

[MS06] Albert Meixner and Daniel J. Sorin. Dynamic verification of memory consistency in cache-coherent multithreaded computer architectures. In *DSN, Proceedings of the International Conference on Dependable Systems and Networks*, pages 73–82, June 2006.

[MS07] Albert Meixner and Daniel J. Sorin. Error detection using dynamic dataflow verification. In *IPACT, Proceedings of the International Conference on Parallel Architectures and Compilation Techniques*, pages 104–118, September 2007.

[MSB+05] Milo M. K. Martin, Daniel J. Sorin, Bradford M. Beckmann, Michael R. Marty, Min Xu, Alaa R. Alameldeen, Kevin E. Moore, Mark D. Hill, and David A. Wood. Multifacet's general execution-driven multiprocessor simulator (GEMS) toolset. *SIGARCH Computer Architecture News*, 33(4):92–99, 2005.

[MSS99] Joao P. Marques-Silva and Karem A. Sakallah. GRASP: a search algorithm for propositional satisfiability. *IEEE Transactions on Computers*, 48(5):506–521, 1999.

[NdPC+04] Jose A. Nacif, Flavio M. de Paula, Claudionor N. Coelho, Fernando C. Sica, Harry Foster, Antonio O. Fernandes, and Diogenes C. da Silva. The Chip is Ready. Am I done? On-chip Verification using Assertion Processors. In *Symposium on Integrated Circuits and System Design*, pages 55–59, September 2004.

[NMGT06] Jun Nakano, Pablo Montesinos, Kourosh Gharachorloo, and Josep Torrellas. ReViveI/O: efficient handling of I/O in highly-available rollback-recovery servers. In *International Symposium on High-Performance Computer Architecture*, pages 203–214, February 2006.

[PBG05] Mukul R. Prasad, Armin Biere, and Aarti Gupta. A survey of recent advances in SAT-based formal verification. *STTT, Journal on Software Tools for Technology Transfer*, 2:156–173, April 2005.

[PF05] Douglas L. Perry and Harry Foster. *Applied Formal Verification: For Digital Circuit Design*. McGraw-Hill Professional, first edition, 2005.

[PG74] Gerald J. Popek and Robert P. Goldberg. Formal requirements for virtualizable third generation architectures. *Communications of the ACM*, 17(7):412–421, 1974.

[Piz04] Andrew Piziali. *Functional verification coverage measurement and analysis*. Springer, first edition, 2004.

[Rot00] Hemant Rotithor. Post-silicon validation methodology for microprocessors. *IEEE Design and Test of Computers*, 17(4):77–88, October 2000.

[RS95] Kavita Ravi and Fabio Somenzi. High-density reachability analysis. In *ICCAD, Proceedings of the International Conference on Computer Aided Design*, pages 154–158, February 1995.

[SC04] Tom Shanley and Bob Colwell. *The Unabridged Pentium 4: IA32 Processor Genealogy*. Addison-Wesley Professional, illustrated edition, 2004.

[SE04] Amitabh Srivastava and Alan Eustace. ATOM: A system for building customized program analysis tools. *ACM SIGPLAN Notices*, 39(4):528–539, April 2004.

[SFH+03] Isic Silas, Igor Frumkin, Eilon Hazan, Ehud Mor, and Genadiy Zobin. System-level validation of the Intel® Pentium® M processor. *Intel Technology Journal*, 07:38–43, May 2003.

[SKC95] Bart Selman, Henry Kautz, and Bram Cohen. Local search strategies for satisfiability testing. In *DIMACS Series in Discrete Mathematics and Theoretical Computer Science*, pages 521–532, 1995.

[SLL02] Jeremy Siek, Lie-Quan Lee, and Andrew Lumsdaine. *The Boost graph library: user guide and reference manual*. Addison-Wesley, 2002. `http://www.boost.org/doc/libs/release/libs/graph`.

[SMHW02] Daniel J. Sorin, Milo M.K. Martin, Mark D. Hill, and David A. Wood. SafetyNet: improving the availability of shared memory multiprocessors with global checkpoint/recovery. In *ISCA, Proceedings of the International Symposium on Computer Architecture*, pages 123–134, 2002.

[SNC+07] Smruti Sarangi, Satish Narayanasamy, Bruce Carneal, Abhishek Tiwari, Brad Calder, and Josep Torrellas. Patching processor design errors with programmable hardware. *IEEE Micro*, 27(1):12–25, 2007.

[Som09] Fabio Somenzi. *CUDD: CU Decision Diagram Package*, 2009. `http://vlsi.Colorado.edu/~fabio/CUDD`.

[SS88] Dennis Shasha and Marc Snir. Efficient and correct execution of parallel programs that share memory. *ACM Transactions on Programming Languages and Systems*, 10(2):282–312, 1988.

[STT06] Smruti Sarangi, Abhishek Tiwari, and Josep Torrellas. Phoenix: Detecting and recovering from permanent processor design bugs with programmable hardware. In *MICRO, Proceedings of the International Symposium on Microarchitecture*, pages 26–37, December 2006.

[SV06] Sean Safarpour and Andreas Veneris. Abstraction and refinement techniques in automated design debugging. In *MTV, Proceedings of the International Workshop on Microprocessor Test and Verification*, pages 88–93, December 2006.

[SVM07] Sean Safarpour, Andreas Veneris, and Hratch Mangassarian. Trace compaction using SAT-based reachability analysis. In *ASP-DAC, Proceedings of the Asian-South Pacific Design Automation Conference*, pages 932–937, January 2007.

[Syn09] Synopsys®Inc. *VCS: Comprehensive RTL Verification Solution*, 2009. `http://www.synopsys.com/tools/verification/functionalverification/pages/vcs.aspx`.

[Syn10a] Synopsys®Inc. *Design Compiler 2010*, 2010. `http://www.synopsys.com/tools/implementation/rtlsynthesis/pages/designcompiler2010-ds.aspx`.

[Syn10b] Synopsys®Inc. *Magellan: Hybrid RTL Formal Verification*, 2010. `http://www.synopsys.com/tools/verification/functionalverification/pages/magellan.aspx`.

[Val07] Theo Valich. AMD delays Phenom 2.4 GHz due to TLB errata. *The Inquirer*, November 2007. `http://www.theinquirer.net/inquirer/news/995/1025995/amd-delays-phenom-ghz-due-tlb`.

[VHR+07] Sriram Vangal, Jason Howard, Gregory Ruhl, Saurabh Dighe, Howard Wilson, James Tschanz, David Finan, Priya Iyer, Arvind Singh, Tiju Jacob, Shailendra Jain, Sriram Venkataraman, Yatin Hoskote, and Nitin Borkar. An 80-tile 1.28TFLOPS network-on-chip in 65nm CMOS. In *ISSCC, Proceedings of the International Solid State Circuits Conference*, pages 5–7, February 2007.

[VR05] Srikanth Vijayaraghavan and Meyyappan Ramanathan. *A practical guide for system Verilog assertions*. Springer, illustrated edition, 2005.

[WB07] Ilya Wagner and Valeria Bertacco. Engineering trust with semantic guardians. In *DATE, Proceedings of Design, Automation and Test in Europe Conference*, pages 743–748, April 2007.

[WB08] Ilya Wagner and Valeria Bertacco. Reversi: Post-silicon validation system for modern microprocessors. In *ICCD, Proceedings of the International Conference on Computer Design*, pages 307–314, October 2008.

[WB09] Ilya Wagner and Valeria Bertacco. Caspar: Hardware patching for multi-core processors. In *DATE, Proceedings of Design, Automation and Test in Europe Conference*, pages 658–663, April 2009.

[WBA06] Ilya Wagner, Valeria Bertacco, and Todd Austin. Shielding against design flaws with field repairable control logic. In *DAC, Proceedings of the Design Automation Conference*, pages 344–347, July 2006.

[WBA07] Ilya Wagner, Valeria Bertacco, and Todd Austin. Microprocessor verification via feedback-adjusted Markov Models. *IEEE Transactions on Computer-Aided Design of Integrated Circuits and Systems*, 26(6):1126–1138, 2007.

[WBA08] Ilya Wagner, Valeria Bertacco, and Todd Austin. Using field-repairable control logic to correct design errors in microprocessors. *IEEE Transactions on Computer-Aided Design of Integrated Circuits and Systems*, 27(2):380–393, 2008.

[WGK90] David A. Wood, Garth A. Gibson, and Randy H. Katz. Verifying a multiprocessor cache controller using random test generation. *IEEE Design and Test of Computers*, 7(4):13–25, 1990.

[WGR05] Bruce Wile, John C. Goss, and Wolfgang Roesner. *Comprehensive functional verification the complete industry cycle*. Morgan Kaufmann, illustrated, annotated edition, 2005.

[WOT+95] Steven Woo, Moriyoshi Ohara, Evan Torrie, Jaswinder Singh, and Anoop Gupta. The SPLASH-2 programs: characterization and methodological considerations. In *ISCA, Proceedings of the International Symposium on Computer Architecture*, pages 24–36, June 1995.

[Yer06] Siva Yerramilli. On the need for convergence between design validation and test. In *ITC, Proceedings of the International Test Conference*, page 14, October 2006.

[YPA06] Jun Yuan, Carl Pixley, and Adnan Aziz. *Constraint-Based Verification*. Springer, first edition, 2006.

Index